Y0-BPU-284

ASTRONOMY
A Popular History

ASTRONOMY
A Popular History

J. Dorschner,

C. Friedemann,

S. Marx

and W. Pfau

Illustrations by G. Löffler

VNR VAN NOSTRAND REINHOLD COMPANY
New York Cincinnati Toronto London Melbourne

Translated by C. S. V. Salt

This book was originally published under the title
ASTRONOMIE heute – Gesicht einer alten Wissenschaft.

Library of Congress Catalog Card Number: 74–5942

ISBN 0-442–22168–1

Printed in Germany (East).

Published in 1975 by Van Nostrand Reinhold Company
A division of Litton Educational Publishing, Inc.
450 West 33rd Street, New York, N.Y. 10001, U.S.A.

Van Nostrand Reinhold Limited
1410 Birchmount Road, Scarborough, Ontario M1P 2E7, Canada

16 15 14 13 12 11 10 9 8 7 6 5 4 3 2 1

Library of Congress Cataloging in Publication Data
Main entry under title:
Astronomy, a popular history.

Translation of ASTRONOMIE heute – Gesicht einer alten Wissenschaft.
1. Astronomy–History. I. Dorschner, J.
II. Löffler, G.
QB15.A8713 520'.9 74–5942
ISBN 0–442–22168–1

Contents

Foreword

Astronomy today—the result of 5,000 years of human endeavors to explain the phenomena and happenings in the starry sky. This is the story of a science from the stage of naive primitive observations and mythical interpretations to the precise research methods of the Space Age.

Even though this book is not intended as an historical description in the scientific sense, the first of its four parts—each of which is complete in itself—does attempt an outline of the history of astronomy. It is especially in the other parts that the real aim of the book is reflected. These describe the development of observational techniques and the knowledge of the Cosmos gained with their assistance. This is not a textbook, however, the aim of the authors being to acquaint the reader with significant astronomical discoveries and findings which are of enduring value and part of the great cultural achievements of mankind. Importance has been attached to a readily understandable presentation of the material.

In the very recent past, there has been a rapid increase in the data available in the astronomical field. For this reason, the scope of the material selected has had to be strictly limited. Much important historical and astrophysical data is provided in compressed form as an appendix to the book, together with a glossary of many technical terms.

In the presentation of the material the authors have considered it important to take account of emerging trends in the development of the science of astronomy so that the reader is prepared for future developments, as it were, and in order that the book does not become prematurely outdated despite the present "explosion" in knowledge. The great discoveries are indeed the result of creative human effort, but a detailed analysis of individual personalities and of the protracted and contradictory process of the acceptance of new ideas would be outside the scope of the book.

The illustrations, in most cases, not only make it easier to understand the text but provide additional details of some significance. On the other hand, they are intended to provide an idea of astronomy and its development without the necessarily detailed study of the chapter in question. To make dry and abstract facts more digestible, use of unconventional diagrams has been made to replace the usual formulae, tables and line drawings.

The authors would like to take this opportunity to thank all those who have played a part in the production of this book and especially the many professional colleagues and institutions in East Germany and other countries who so willingly provided them with pictorial material. The numerous reproductions are the work of the Film and Picture Center of Friedrich Schiller University, Jena, G. Schörlitz providing valuable assistance in this connection. Finally, thanks are due to the publishers for the great understanding and interest shown in this project.

Johann Dorschner   Christian Friedemann   Siegfried Marx   Werner Pfau

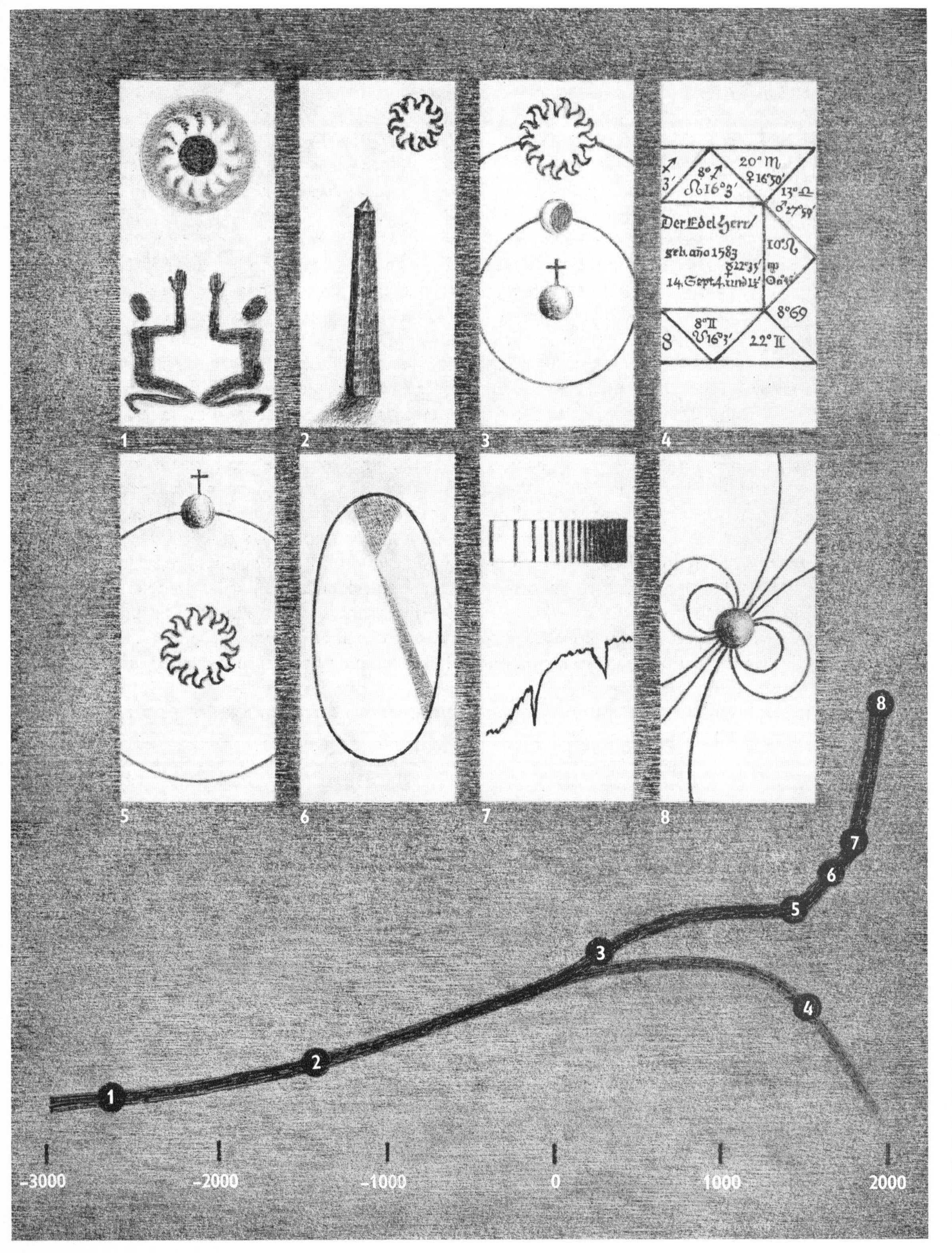

1 In the early days of astronomy, there were close links between it and astrology. People could not explain the phenomena in the sky and were even frightened by some of them, such as an eclipse of the Sun. This partly explains why they worshipped celestial bodies (1). It was not long before they learned to utilize the events taking place in the sky for their everyday life, e.g., for reckoning time (2). Observations, in particular, led to an increasing knowledge of what was happening in the sky. This is symbolically expressed by the slight rise in the lower curve between 3000 B.C. and the year 0. The graphic presentation of this development is essentially only of a symbolic character, of course. Soon after the year 0, the Ptolemaic model of the planetary system provided a theoretical interpretation of the motion of the planets (3). This was followed by a clear slackening in the development of scientific astronomy until the end of the Middle Ages. With the development of the Copernican model of the planetary system (5), a new epoch began in which the development of astronomy made rapid progress, this being due to Kepler's discovery of the laws of planetary motion (6) and Newton's law of gravitation. In the Nineteenth Century, a new field of astronomical activity emerged in the shape of astrophysics, based on the spectroscopic work (7) of Fraunhofer, Kirchhoff and Bunsen. The result of this was that astronomy advanced even more rapidly. At the present time, astrophysics is the branch of astronomy in which the most work is being done, as symbolized by a cosmic magnetic field (8). In addition to the visual spectral range, astronomy now obtains information from radio frequency radiation, infrared radiation and shortwave X-ray and gamma radiation as well. Up to recent times, astrology (4), using a

Chapter I
The evolution of modern astronomy

pseudoscientific fund of knowledge, had numerous followers. As the result of scientific enlightenment it now has little influence.

A great number of astronomical facts now seem self-evident. The length of a year is the time taken by the Earth to move around the Sun. The length of a day is the time taken by the Earth to rotate once around its axis. You can find out where you are on the Earth from the position of the stars in the sky. But the exact time taken by the Earth to move around the Sun cannot be expressed by a whole number of days since it amounts to 365.2425 days. Since the alternation of day and night is a convenient basis for measuring time, however, the calendar in general use is based on whole numbers, the fractions of days over being added together and expressed as an extra day in leap years. This shows that even the elementary problem of reckoning time is not such a straightforward matter.
There are many reasons in astronomy for determining the time taken by the Earth to move around the Sun and it has been discovered that this time varies. As a result, different years cover different lengths of time. There are similar surprises in store for us when we take a look at the stars in the sky. From our everyday experience, we assume that the sky at night does not change, apart from daily and seasonal movements and the courses of the planets. But in 2,000 years' time the Pole Star will not be so close to the North Pole as it is today and in 100,000 years—not a long time in the history of the world—the constellation of Ursa Major (also known as the Great Bear) will not be seen as it is today. However, astronomy is not only the basis of the reckoning of time and position but is also concerned with the material objects and processes outside the Earth. Its area of research, unlike other branches of science, is consequently very far away and can scarcely be influenced in the experimental sense. On the other hand, Space offers physical conditions which can hardly be obtained in laboratories on the Earth. Other branches of science, such as physics, can thus profit from the findings of astronomy. Advances in astronomy, in turn, depend on the level of other sciences and of technology as well. As a consequence, the development of astronomy may only be understood when account is taken of the social factors influencing it and of the close interrelations with other scientific disciplines.
Let us return to the things which today are regarded as self-evident. Even these simple astronomical facts were not always understood by mankind but are the result of protracted and creative work.

Search for signs and portents

Practical social needs were what led Man to turn his attention to the stars. A method of reckoning time was needed for deciding the most favorable dates for sowing and harvesting or hunting and fishing, i.e., for ensuring the basis of life. It was soon realized that the periodical reappearance of certain phenomena in the night sky could easily be used for this, and thus a calendar system was worked out on the basis of these phenomena. Another reason why attention was paid to stars was the need for a system of guidance in the desert or on the high seas, for instance. Here, too, practical requirements were the stimulus for astronomical observations.
Apart from this, the functions of astronomy during the early stages of its development featured mythical and religious aspects to a marked degree. Since the Moon, the planets and above all the Sun were honored as deities, attempts were made to draw conclusions from their path across the heavens concerning what was in store for mankind. It was this which formed the roots of astrology. It is to be noted that in the early days of celestial observation astronomy and astrology formed an indivisible whole.
For the reckoning of time and orientation by the stars, it was necessary to observe the apparent positions of the stars and their apparent movements. In general, the early astronomy of the civilized peoples of Antiquity was relatively well-developed. Evidence exists of the astronomical activities of the Babylonians, Egyptians, Chinese, Indians and Mayas. The same level of development was not attained everywhere and there was also reciprocal influence to some extent, but all these ancient civilized peoples made some contribution to the collection of astronomical data by observation.
It is difficult to say exactly when astronomy became a science. Although this book does not offer a detailed history of astronomy, a few remarks on this aspect are nevertheless justified. All peoples were aware, of course, that the Sun rises in the East and sets in the West and that in the warm season of the year it shines longer and rises higher above the horizon than in the cold period. All peoples divided time into days, months and years which were often of different length. The phases of the Moon, the varying degrees of brightness of the stars and the daily movement of the starry sky did not escape the notice of observers. However, it is only possible to talk of astronomy as such after the transition from naive

observation to the systematic compilation of observational data, with the preparation, for instance, of lists of constellations arranged according to their position in the sky or according to the months in which they again become visible for the first time. It is this which marks the emergence of systematically recorded knowledge which could be handed down to later generations for further use.

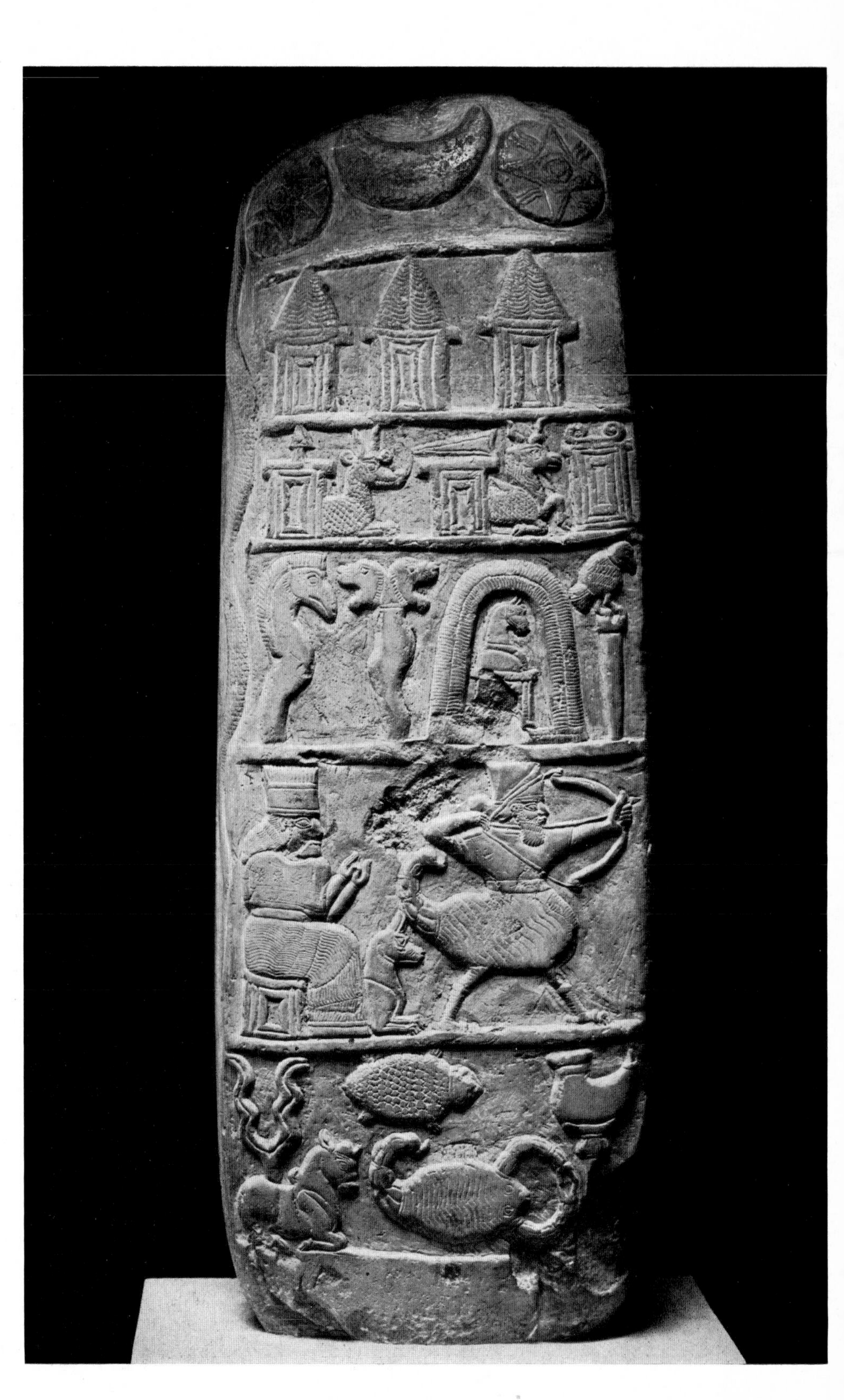

The Babylonian legacy

The lists of stars compiled by the Babylonians were constantly improved in the course of the centuries and by about 700 B.C. a proper catalog of stars was available with accurate information on the relative position, appearance, culmination and disappearance of the stars. It also contained tables of the lengths of shadows, the duration of the night, the visibility of the Moon and details of the paths of the Sun, the Moon and the planets. This was a first step towards scientific astronomy.

The Babylonian sphere of civilization has been selected because it is only possible here to provide an outline of the history of astronomy as far as Copernicus and to trace the consequences of its development. The astronomy of the early stages of modern times was derived from Greek astronomy which, in turn, was in touch with the Babylonian and Egyptian civilizations. Thus the study of the stars by the Babylonians was the beginning of a development which has continued down to the European astronomy of our period. At the same time, though, we have also inherited astrology from the Babylonians, a legacy which is not so positive in its implications.

The astronomers of Babylon collected a great deal of data as the result of their observations. They acquired a wider and more profound knowledge of celestial phenomena and were ultimately able to progress from the simple recording to the prediction of celestial movements. The time taken to complete their orbits by the planets visible to the naked eye and the rules for the long-term prediction of eclipses were known to Babylonian astronomy, which reached its peak in the last five or six centuries B.C.

The achievements of Egyptian astronomy are more modest. At a very early date—probably in the Fourth Millennium B.C.—the Egyptians took the Sun as the basis of their calendar and, unlike other civilized peoples, did not use the phases of the Moon for calculating time. They also introduced the four-year cycle of leap years more than two thousand years ago.

2 In Babylon, long before the birth of Christ, astronomical symbols were also used on richly decorated boundary stones, as shown here at the top. These boundary stones carried symbols of various deities as well, the purpose of this being to conjure up the curse of the gods depicted on anyone who dared to remove the boundary stone or to infringe the rights of its owner in other ways.

3 The picture shows the upper part of the stele of King Asarhaddon of Assyria who ruled from 680 to 669 B.C. after a revolt against his father. This historically valuable relief was found in Sam'al (Sandshirli) in Northern Syria. In the upper right corner there are depictions and symbols of deities, these being closely associated with astrological procedures.

The contributions of Hellas

Greek astronomers were able to utilize the great experience of the Babylonians and Egyptians and played a great role in the further development of this science. However, only the really new contributions made by this civilization to the advance of astronomy will be considered here. In contrast to earlier epochs, the Greeks were more interested in the causes of the phenomena than in the phenomena as such. They accordingly made the first attempts to base the movements of the stars on a mathematical system, i.e., to work out a model of the mechanics of the movements of the planets and of the "fixed stars attached to the celestial sphere." This period reached its climax with the geocentric system proposed by Hipparchus and mathematically expressed by Ptolemy. According to this, the Earth is the center of the universe and all the other planets within the sphere of the fixed stars move around the Earth in complicated orbits.

A noteworthy fact is that a heliocentric system, too, was known to the Greeks (e.g., that of Aristarchus of Samos). An important aspect in the acceptance of the geocentric interpretation of the world in Antiquity was that it complied with the requirements of Aristotelian philosophy for perfect figures and movements.

Greek astronomy reached its zenith with Ptolemy's famous *Syntaxis mathematica* (The Mathematical Collection), known as the *Almagest*. This important astronomical work of Antiquity contains not only a description of the geocentric theory of the universe but also a review of Greek astronomy in general.

In the main, it is the Arabs to whom credit must be given for the preservation of the tremendous knowledge of the Greeks. They perfected the mathematical aids used in astronomy, improved the instruments inherited from Antiquity and added new findings to the fund of knowledge already accumulated. So it was that astronomy enjoyed a new Golden Age.

4 This lion with astronomical symbols is part of a monument which stood on the mountain of Nimrud Dag and depicted the horoscope of the coronation of King Antiochus I of Commagene. It is believed that the coronation took place on July 7, 62 B.C. when the Sun was in the constellation of the Lion. On the breast of the beast there can be seen the crescent of the Moon, enclosing Regulus, the brightest star in the constellation of the Lion.

5 In the grave of Rameses VII, dating from the Twelfth Century B.C., there are pictures of an astronomical method of reckoning time. Over each man, seated in a row, there are twelve lines representing the twelve parts of the night. The vertical lines, on which stars are marked, designate positions of the stars. In each line a star is mentioned which can be seen in a certain position at this time. The text describes the position as follows: "Above the left ear" or "Above the left shoulder" and so on. The row consists of 24 figures in all, there being one figure each for the 1st and 16th day of every month. This method of reckoning time soon gave way to better systems.

Revolution in the Renaissance

A revolution in astronomical thinking then occurred at the beginning of modern times. The endeavors of the progressive middle class to broaden their educational horizon and to overcome the scholastic approach of medieval learning resulted in a thorough rethinking of astronomical problems.

In the age of the great discoveries, new practical needs also emerged as a consequence of social development and the upswing in production and commerce.

6 On the ceiling of the temple of Dendera, the famous city of the Ptolemaic-Roman period in Upper Egypt, there is this "round Zodiac." All twelve signs of the Zodiac may be seen in the central part of the picture. On the right there are the Pisces, with Aries and Taurus below them. To the left of center, there is Gemini, surmounting Cancer and Leo. Above Gemini, Virgo, Libra and Scorpio can be seen. Sagittarius, Goatfish and Aquarius are shown above the center. Goatfish is Babylonian and is equivalent to Capricornus. Around the circumference, the decans are depicted as serpents and other creatures and as striding men. Originally, the decans were simply constellations by which the time at night could be ascertained. They later appeared in astrological literature as deities who determined human fate.

7 The "Stone of the Sun," a calendar stone, which was completed in 1479 after more than fifty years of work, is one of the most magnificent monuments of Aztec art. It is a great block of basalt, 25 tons in weight and 3.6 meters (141.7 inches) in diameter. In the center, there is the countenance of the Sun God of the Aztecs. The four squares around the head of the Ruler of the World probably represent the four seasons. The circle surrounding these is divided into 20 panels, each of which contains a sculpture symbolizing the 20 days of the Aztec calendar. A year consists of 18 months of 20 days each. This amounts to 360 days to which another five—marked as five large points within the Circle of the Days—are added. On each of these Sacrificial Days, an Aztec of high rank was sacrificed to maintain the good will of the Sun God.

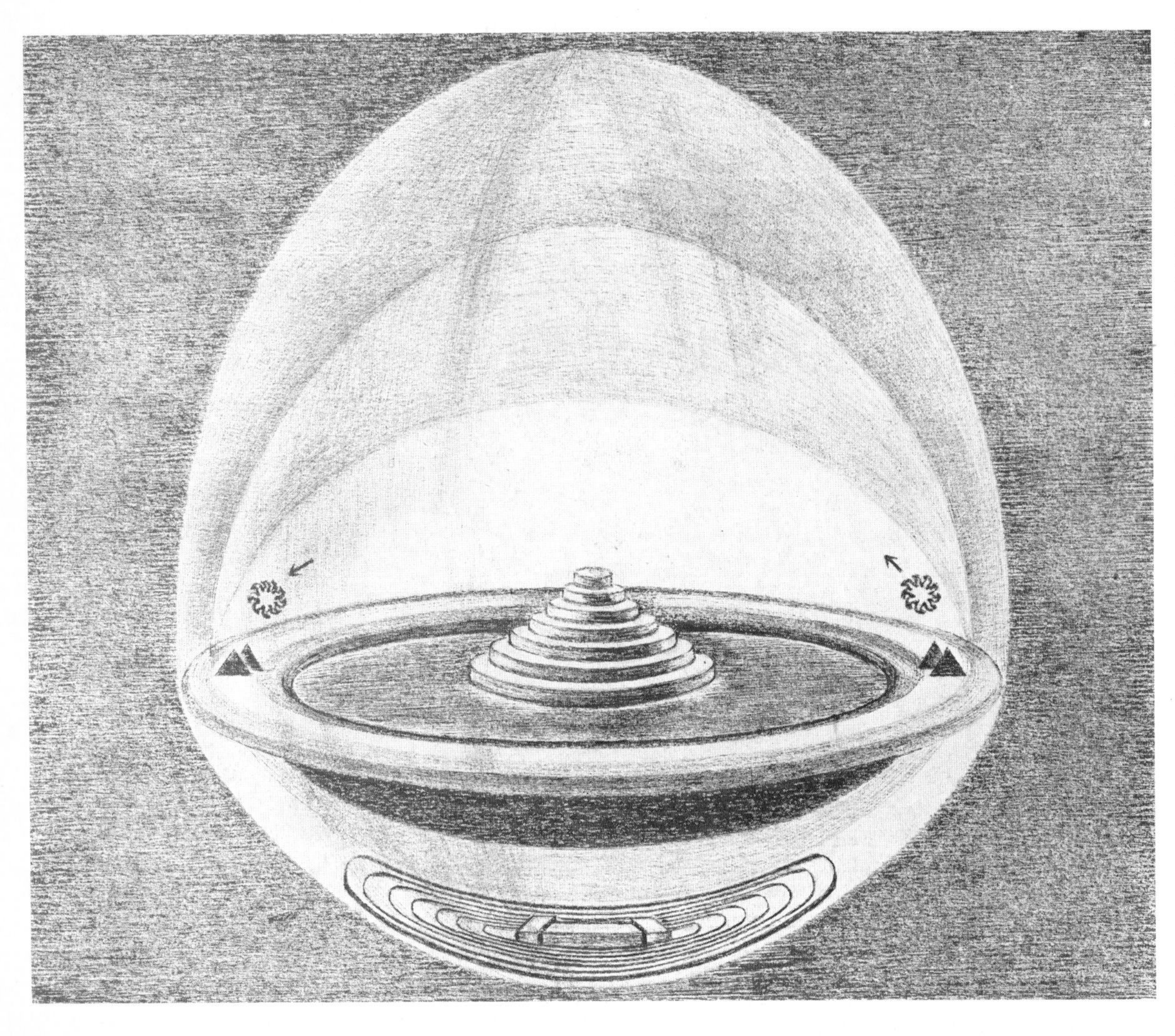

8 The ideas of the Babylonians on the arrangement of the world can be reconstructed from written records which have survived. The Upper World—represented by the stepped tower—floats on the terrestrial ocean with the first, second and third heavens above it. Below the terrestrial ocean there is the Underworld with the Palace of the Realm of the Dead within its seven walls. The terrestrial ocean is enclosed within the Causeway of the Sky on which the Mountains of the Dawn and Sunset are to be seen.

It was no accident that the problems of the universe were examined from a new standpoint at the beginning of the Sixteenth Century with the stimulus of the critical spirit of the Renaissance. On the basis of the old observations, Nicholas Copernicus (1473—1543) developed his heliocentric theory, by which the Earth and thus Man were banished from the center of creation. Although there were shortcomings in the Copernican system due to the retention of orbital circles and despite the fact that it was not at all superior to the Ptolemaic system in practical terms, it did initiate "the emancipation of astronomy from theology" (Engels) and represents the beginning of a new era in astronomical thought. Seen in this way, it really must be termed the "Copernican revolution." Nevertheless, it is quite logical that the Copernican system had to overcome prejudice not only on the part of astronomers but also of philosophers and theologians as well before it could meet with general recognition.

Of fundamental importance for the post-Copernican advance of astronomy was the development of the natural sciences and the improvement of observational techniques, this being closely associated with the development of social relations. The identification of the laws of planetary motion by Kepler was a brilliant confirmation of the heliocentric theory. The physical foundation for Kepler's laws was the discovery of the law of universal gravitation by Newton. Newton's mechanics and his law of gravitation then formed the foundation-stone for the magnificent structure of the celestial mechanics of the

9 The scientific publication has always been one of the most important information media and aids for successful research work. During the Middle Ages and at the beginning of modern times, scientific papers were usually published in Latin and, for this reason alone, could only be read by a small number of people. Among others, this is also true of the principal work of Copernicus which was certainly one of the most important sources of astronomical information for Kepler. The illustration shown here is taken from Kepler's personal copy, and the marginal notes are in his handwriting.

REVOLVTIONVM LIB. V. 142

Quibus modis errantium motus proprij appareant inæquales. Cap. IIII.

Voniam uero motus eorũ ſecundũ lõgitudinẽ proprij eundem ferè modum habẽt, excepto Mercurio, qui uidetur ab illis differre. Quamobrẽ de illis quatuor cõiunctim tractabitur. Mercurio alius deputatus eſt locus. Quòd igitur priſci unũ motum in duobus eccentris (ut recenſitũ eſt) poſuerunt, nos duos eſſe motus cenſemus æquales, qbus inæqualitas apparentiæ componitur, ſiue ꝑ eccentri eccentrũ, ſiue ꝑ epicycli epicycliũ, ſiue etiam mixtim ꝑ eccẽtrepicyclũ, quæ eandẽ poſſunt inæqualitatem efficere, uti ſuperius circa Solem & Lunã demõſtrauimus. Sit igitur eccentrus A B circulꝰ circa C cẽtrum, dimetiens A C B medij loci Solis per ſummã ac infimã abſida planetæ, in qua centrũ orbis terreni ſit D, facto ꝗ in ſumma abſide A, Diſtantiæ aũt tertiæ ꝑtis C D, deſcribatur epicycliũ B F, in cuius perigæo quod ſit F, planeta cõſtituatur. Sit aũt motus epicyclij per A B eccentrũ in cõſequentia. Planetæ uero in circũferẽtia epicyclij ſuperiori ſimiliter in

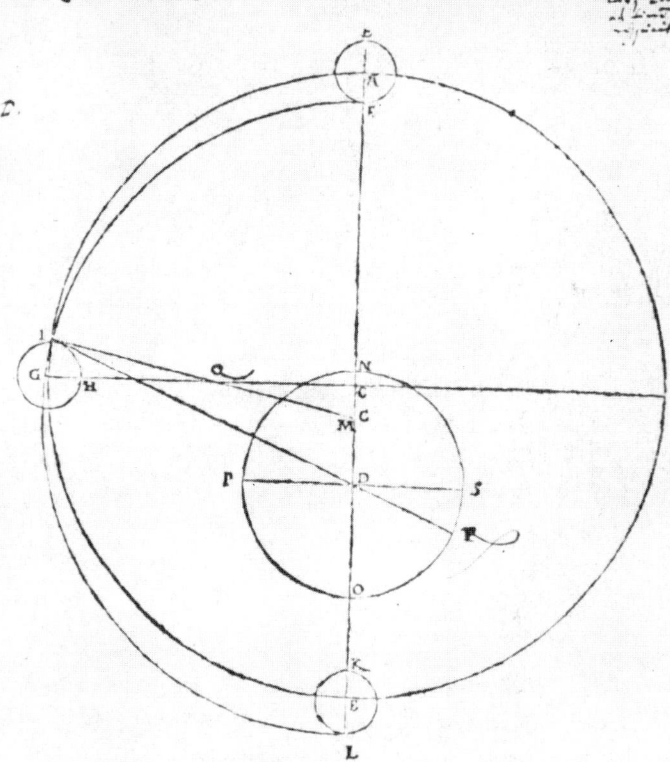

N ij conſequen

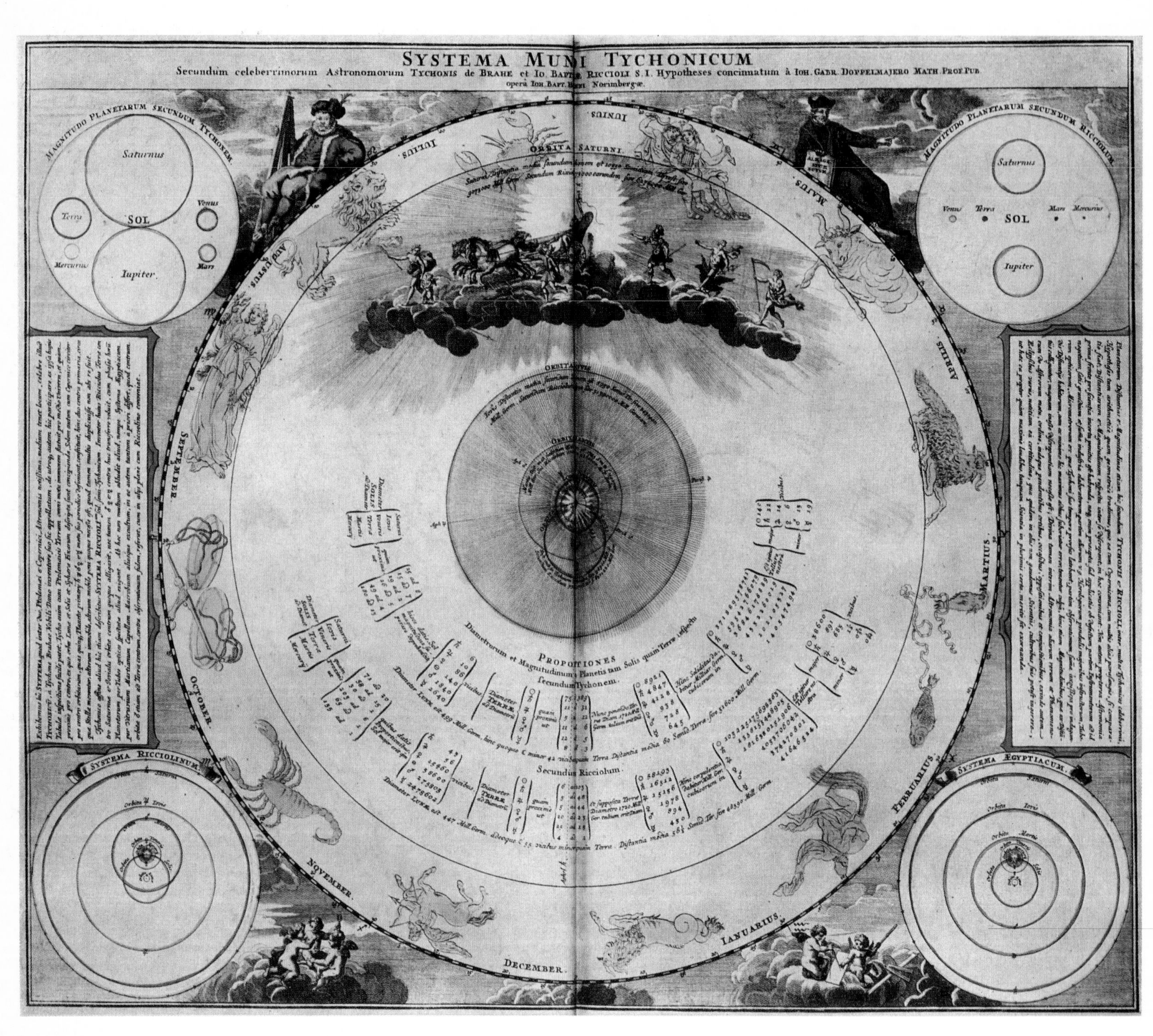

10 In the model of the planetary system by Tycho Brahe the Moon and the Sun move on circular orbits around the Earth. All the other planets move around the Sun. Other "combined" systems of this kind are the "Systema Ricciolinum" and the "Systema Aegyptiacum." In the former, the Moon, the Sun and the planets Jupiter and Saturn orbit the Earth, while Mercury, Venus and Mars move around the Sun. In the "Systema Aegyptiacum," the Moon, the Sun and the planets Jupiter and Saturn circle around the Earth, while only Mercury and Venus circumnavigate the Sun. These two models are shown in the corners at the bottom. In the upper corners are depicted Brahe's and Riccioli's conceptions of the magnitudes of the planets in relation to the Sun. Riccioli was the more accurate of the two.

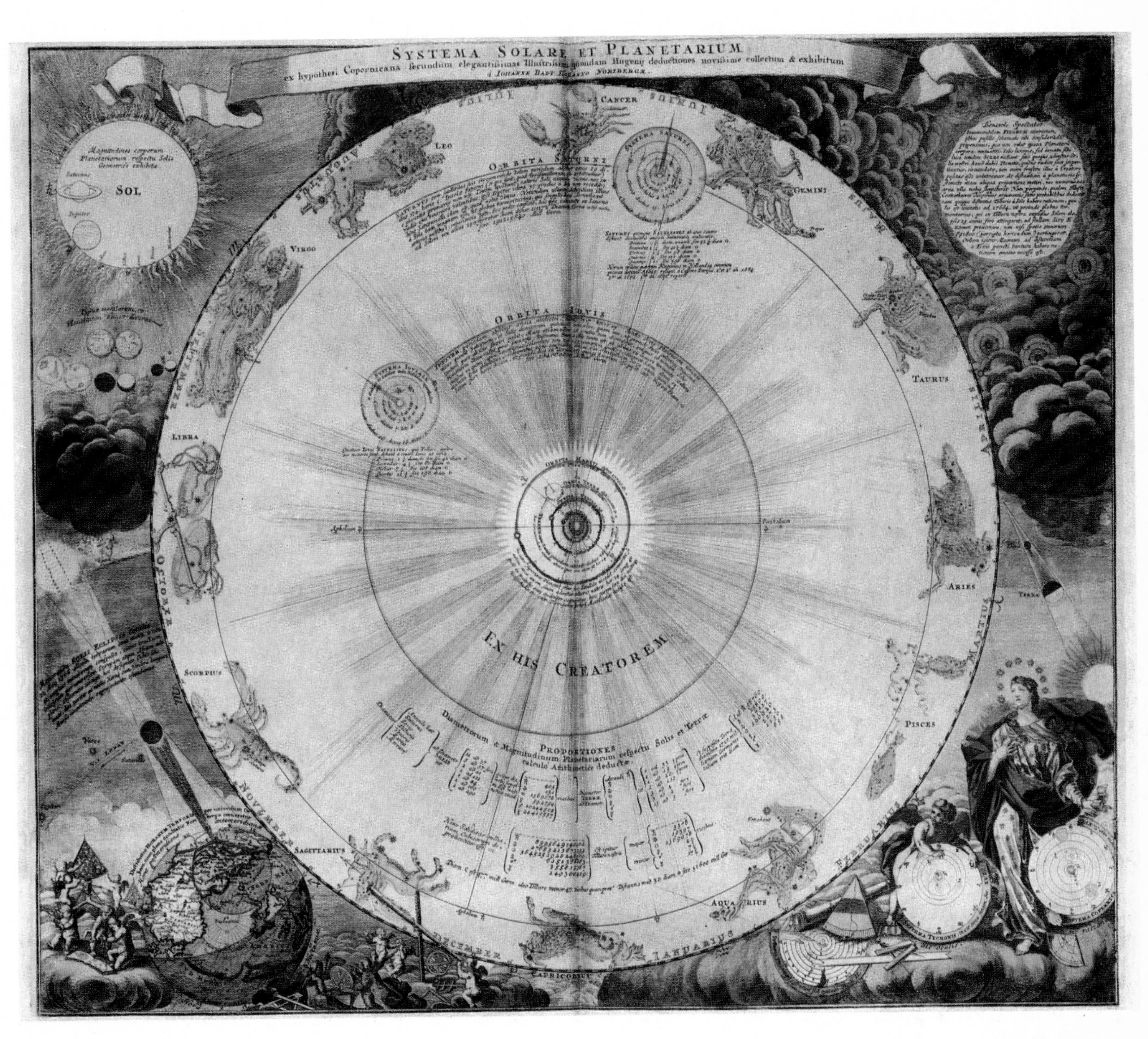

11 This diagram of the Copernican model of the planetary system contains all the planets and planetary satellites known at the time it was drawn. The planets are those which can be observed with the naked eye. Apart from the Earth's Moon, the four satellites of Jupiter discovered by Galileo in 1610 are included. In 1655, Huygens discovered the largest satellite of Saturn, Titan. Another four satellites of Saturn were found by Cassini (Japetus 1671, Rhea 1672, Dione and Tethys 1684). The entire planetary system is enclosed here in the circle of the twelve symbols of the Zodiac. Other astronomical phenomena, such as the relative sizes of the planets and the occurrence of eclipses of the Sun and Moon, are used as a frame for the picture.

Eighteenth Century, the prototype of an exact natural science. A significant achievement of this branch of science was the theoretical prediction and subsequent discovery (in 1846) of the planet Neptune which had remained unknown until that time.
When the historical development of astronomy is traced from its primitive beginnings to the state described in this book, it is apparent that this science has not only made a contribution to satisfying the practical needs of mankind but has also played a decisive part in the development of a scientific view of the world. It is precisely this latter point which explains why modern astronomy has always been and still is at the center of philosophical discussion.

Paris, le 18 Septembre 1846.

Monsieur,

J'ai lu avec beaucoup d'intérêt et d'attention la réduction des Observations de Roemer, dont vous avez bien voulu m'envoyer un exemplaire. La parfaite lucidité de vos explications, la complète rigueur des résultats que vous nous donnez, sont au niveau de ce que nous devions attendre d'un aussi habile Astronome. Plus tard, Monsieur, je vous demanderai la permission de revenir sur plusieurs points qui m'ont intéressé, et en particulier sur les Observations de Mercure qui y sont renfermées.
Aujourd'hui, je voudrais obtenir de l'infatigable observateur qu'il voulût bien consacrer quelques instants à l'examen d'une région du Ciel, où il peut rester une Planète à découvrir. C'est la

A Monsieur,
Monsieur J. G. Galle,
Astronome à l'Observatoire royal de Berlin
à Berlin.

FRANCO

théorie d'Uranus qui m'a conduit à ce résultat. Il va paraître un extrait de mes recherches dans les Astr. Nachr. J'aurais donc pu, Monsieur, me dispenser de vous en écrire, si je n'avais eu à remplir le devoir de vous remercier pour l'intéressant ouvrage que vous m'avez adressé.

Vous verrez, Monsieur, que je démontre qu'on ne peut satisfaire aux observations d'Uranus qu'en introduisant l'action d'une nouvelle Planète, jusqu'ici inconnue : et ce qui est remarquable, il n'y a dans l'écliptique qu'une seule position qui puisse être attribuée à cette Planète perturbatrice. Voici les éléments de l'orbite que j'assigne à cet astre :

Demi grand axe de l'orbite ... 36,154
Durée de la révolution sidérale ... 217,387
Excentricité ... 0,10761
Longitude du Périhélie ... 284° 45'
Longitude moyenne 1er Janvier 1847 ... 318° 47'
Masse ... 1/9300
Longitude Héliocentrique vraie au 1er Janvier 1847 ... 326° 32'
Distance au Soleil ... 33,06

La position actuelle de cet astre montre que nous sommes actuellement, et que nous serons encore, pendant plusieurs mois, dans des conditions favorables pour la découvrir.

D'ailleurs, la grandeur de sa masse permet de conclure que la grandeur de son diamètre apparent est de plus de 3" sexagésimales. Ce diamètre est tout-à-fait de nature à être distingué, dans les bonnes lunettes, du diamètre factice que diverses aberrations donnent aux étoiles.

Recevez, Monsieur, l'assurance de la haute considération de votre dévoué serviteur,

U. J. Le Verrier

Veuillez faire agréer à Mr. Encke, bien que je n'aie pas l'honneur d'être connu de lui, l'hommage de mon profond respect.

12 On September 18, 1846, Leverrier wrote to his colleague Galle in Berlin and advised him of the result of his theoretical considerations and calculations. This historically significant letter reads as follows: "Today I would like to ask the indefatigable observer to devote a few moments to exploring a region in the sky where it may be possible to discover a planet. It is the theory of Uranus which has led me to this result." Galle received this letter within five days, on the morning of September 23, 1846, and found the new planet, from the details given by Leverrier, during the following night. The remarks about the ephemerides of the predicted planet which Leverrier makes in the letter refer to January 1, 1847.

Paris, 1er Octobre 1846.

Monsieur,

Je vous remercie cordialement de l'empressement que vous avez mis à m'instruire de vos observations du 23 et du 24 Septembre. Grâce à vous, nous voilà définitivement en possession de ce nouveau monde. Le plaisir que j'ai éprouvé de voir que vous l'avez rencontré, à moins d'un degré de la position que j'avais donnée, est un peu troublé par l'idée qu'en vous écrivant plus tôt, il y a quatre mois, nous aurions obtenu dès lors le résultat que nous venons d'atteindre.

Je communiquerai votre lettre, lundi prochain, à l'Académie des Sciences.

Permettez-moi d'espérer que nous continuerons fréquemment une correspondance qui commence sous d'aussi heureux auspices.

Je suis avec une haute considération, Monsieur, votre très dévoué serviteur

U. J. Le Verrier

Soyez assez bon pour m'écrire rue St. Thomas d'Enfer no. 5. Je n'appartiens pas à l'Observatoire, où vous m'aviez adressé votre lettre.

Le Bureau des Longitudes s'est prononcé ici pour Neptune. Le signe un trident. Le nom de

PARIS 1 OCT.

A Monsieur
Monsieur Galle
Astronome à l'Observatoire
de Berlin.

13 When a previously unknown celestial body is found, the person discovering it has the right to give it a name. In 1846, a new planet was discovered twice, as it were. Leverrier postulated its existence on the basis of theoretical considerations and, as a consequence of these, Galle was the first to find it in the sky. Galle left it to Leverrier alone to decide on a name but in a letter to him he once suggested the name "Janus" for the new planet. However, this name was rejected by Leverrier. F. D. Arago, the director of the Paris Observatory, gave the planet the name "Leverrier," this being done with Leverrier's agreement or without any objection being made by him. This met with the disapproval of all astronomers, although there was great enthusiasm about the "theoretical discovery" of a celestial body. The dispute was ended by a proposal

Chapter II
The sharpened vision of mankind

from the "Bureau des Longitudes" and, as a result, the new planet of 1846 has since been known as Neptune, as mentioned in the footnote of the letter from Leverrier to Galle.

Modern astronomy possesses a wealth of knowledge concerning the celestial bodies despite the cosmic distances which generally separate the astronomer from the subjects of his studies. Cosmic space and time are associated with magnitudes quite different to those normally in use on the Earth. Important information on the distance and motion of the celestial bodies, the data on their physical characteristics and the knowledge of their age and development are based solely on an analysis of the light and more often of the radiation that reaches us after traversing distances so great as to be inconceivable for us. Observation is the basis of all astronomical knowledge, but only when it is employed in close association with theoretical considerations can it lead to findings capable of further development. Only an association such as this is able to produce a reasoned interpretation of the observations and permit systematic search for and the ultimate penetration of natural laws.

The speed with which observation techniques and astronomical knowledge have developed is closely correlated with the general development of the natural sciences and technology in human society. Both are characterized by progress which is initially slow but subsequently accelerates, finally attaining an almost breath-taking pace in our own era.

For thousands of years, the happenings in the sky were noted with simple sighting devices of stone, wood or metal. Less than four hundred years have passed since the invention of the telescope, yet the astronomer is already working with automatically controlled instruments of giant size. Less than two decades ago we witnessed the start of artificial satellites only the size of a ball, but in the meantime men have landed on the Moon and complicated equipment has been stationed in Space.

Stone Age observatories

From an earlier period of civilization mighty stone monuments survive as testimony to a Stone Age astronomy. In this early stage of human history, the observation of the sky served for determining the times of the vernal and autumnal equinoxes, of the solstices and also of certain positions of the Moon — and thus for fixing the time of religious festivals and sacrifices. The Stone Age "observatories" were places of worship and many of them suffered destruction as the result of ignorance of their function or lack of interest. More than fifty of these great stone structures still survive in Northern France, the British Isles and in the north of the Federal Republic of Germany. A well-known site is the "Visbeker Bräutigam" on Ahlhorn Heath, southwest of Bremen. A "road," 100 yards long and lined by fifty blocks of stone on each side, was built here 3,500 or 4,000 years ago to ascertain, by its east-west orientation, the point in time at which the equinox occurred. It is only on the 21st of March and 23rd of September of our calendar that the Sun rises exactly in the East or sets in the West. The "Visbeker Bräutigam" is aligned precisely along these points.

On St. Kilda, an island in the Atlantic west of the Hebrides, a stone grave 55 feet long points directly at the rising Sun at the time of the winter solstice. Here was awaited the day on which the Sun, from its lowest position, would begin to rise higher in the heavens again—a flash of brightness in the long and stormy winter of this zone.

The most imposing Stone Age monument with astronomical associations is Stonehenge, situated in the county of Wiltshire about 80 miles southwest of London. It is now believed that this sanctuary was built in three stages, the oldest of which dates back before 1900 B.C. At that time it was initially only a circular bank enclosed by a ditch of more than 300 feet in diameter, within which there was a ring of 56 small pits. These are known as the Aubrey holes, after the English antiquarian who prepared the first plan of Stonehenge in the middle of the Seventeenth Century. Toward the northeast, the bank opens into a long avenue, the direction of which is marked by several great stones. It was only at later dates that the great uprights and lintels were erected which constitute the characteristic picture of Stonehenge. The sanctuary was probably completed by about 1600 B.C.

As early as the middle of the Eighteenth Century, it had been pointed out that the stones marking the northeast entrance in the bank indicate the position of the rising Sun at the time of the summer solstice, our 21st of June. Closer examination of the layout reveals that a whole series of other important positions of the Sun and Moon are associated with the grouping of the stones. In comparison with other Stone Age observation centers which have survived, Stonehenge may almost be regarded as a "multipurpose" site. The methods of probability theory have been used to check whether the arrangement of the stones might still be coincidental and that their apparent relation with the position of the stars might

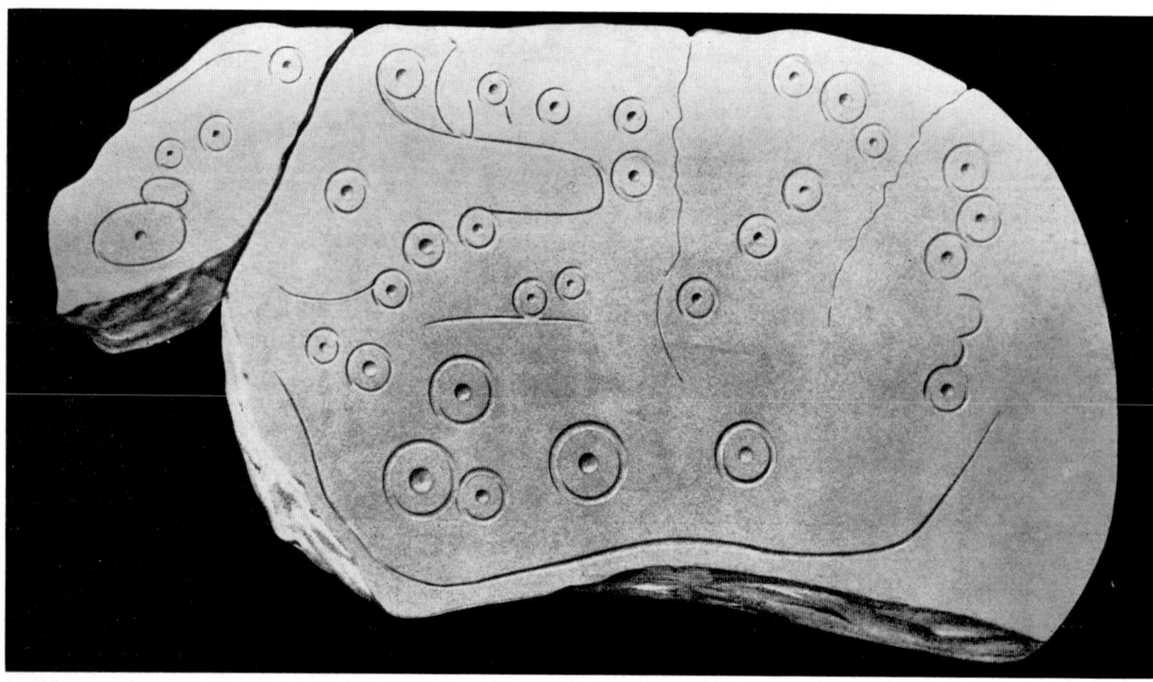

14 This Stone Age celestial chart dates from the Third Millennium B.C. and shows the sky in the area of Scorpio and Sagittarius. The rock drawing was found south of the Caucasus, at Lake Sevan.

15 Stonehenge, the Stone Age cult center in Southern England, is now a tourist attraction. In the medium distance on the photograph, an encircling ditch with the Aubrey holes in front of it may be identified, the monumental main part of the site occupying the foreground. Not only the astronomical knowledge of its Stone Age builders but also their obvious technical skill excite our admiration. The massive blocks of stone weigh up to 50 tons and come from a quarry more than 120 miles away.

only be a subjective interpretation. It appeared, however, that the chances of a coincidence are a million to one and it can consequently be assumed that astronomical reasons played a part in the building of Stonehenge almost 4,000 years ago. The latest investigations have even made use of computer techniques to establish every possible combination of the positions of the stones and of the Aubrey holes with reference to their astronomical significance. As a result, it is now believed that the people of the Stone Age at Stonehenge were even able to predict the occurrence of eclipses of the Sun and Moon. This far-reaching hypothesis has met with enthusiastic support and also with serious criticism. Nevertheless, it is already certain that as early as the Stone Age there must have been considerable knowledge of celestial occurrences. It is to be noted once again that at this stage of civilization the knowledge acquired was employed exclusively for purposes of a religious nature.

Gnomons and armillary spheres

The spiritual source of astronomy as we know it is in the Babylonian Empire. Cuneiform texts provide evidence of activities in astronomical matters as early as the Second Millennium B.C. The basic observations required for the preparation of a calendar were the first step. Regular observations of the Sun, the Moon and of the planets—Venus, in particular—were already being made in the realm of the early Babylonians, the motive here being their belief in the divinity of the celestial bodies with a consequent interest in astrological interpretations. In this sense, the clay tablets which have survived reveal a steadily increasing knowledge over many centuries of astronomy, an increasingly more profound understanding of the complicated movements of the planets. In the last centuries B.C., men even succeeded in applying arithmetical methods for calculating and predicting the behavior of planets, eclipses and even the complicated course of the Moon.

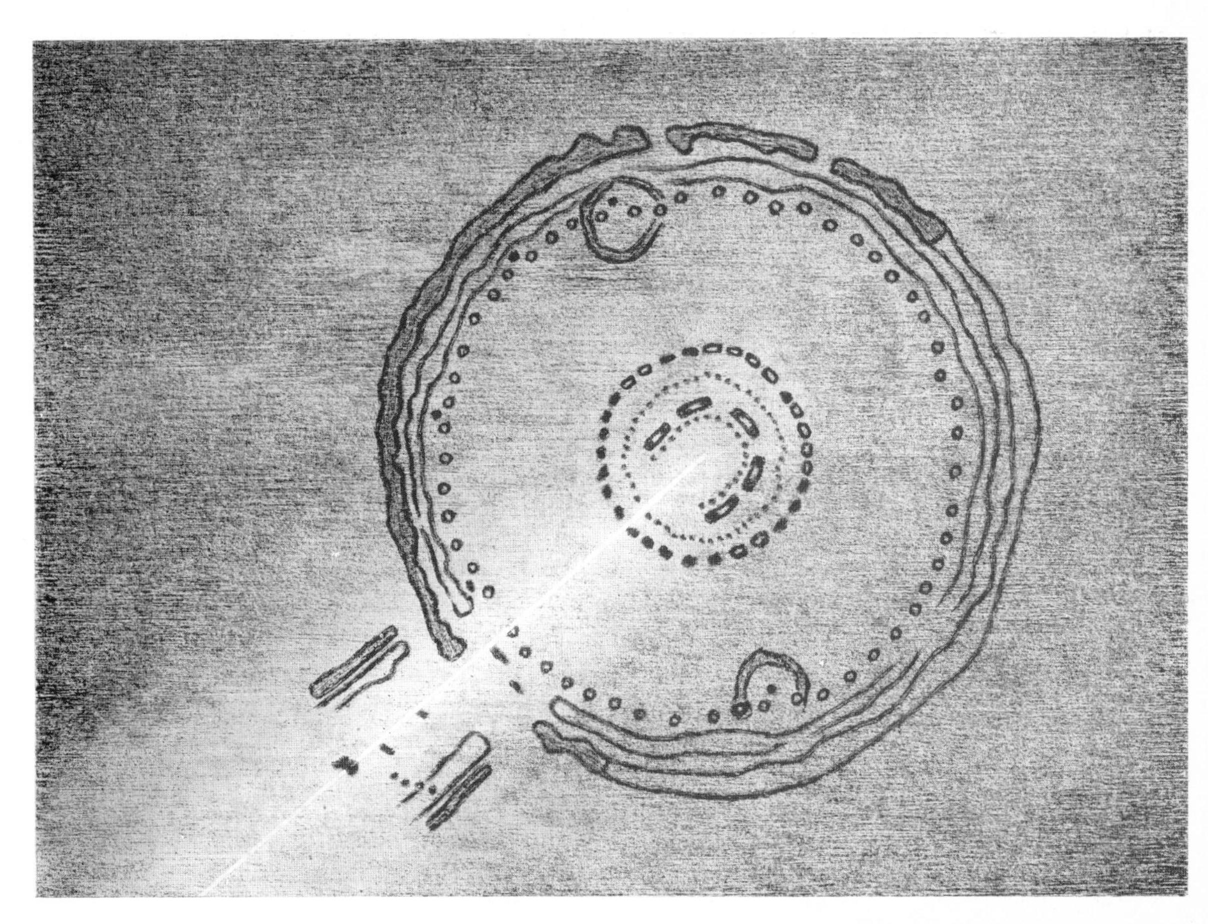

16 From the parts which still survive, it is possible to reconstruct the entire layout of Stonehenge. Right on the outside there is the ditch and mound, enclosing the Aubrey holes, two circles of stones and finally, in the center, the five great trilithons known as the "bluestone horseshoe" from their material and shape. Towards the northeast, which is here at the bottom left as compared with the preceding illustration, there is an opening in the mound in the direction of the most northerly point of sunrise. Two artificial mounds at the northern and southern inner edge of the ditch are indicated.

On the other hand, little can be said about the astronomical instruments of this period. It is known that a rod casting a shadow on a flat surface—the gnomon—was used to indicate the time of day from the length of its shadow. This device was subsequently adopted by the Greeks and has survived down to the present day as the pointer on sundials. The systematic observations of the positions of the Moon and the planets were based, however, purely on estimates of their positions relative to brighter fixed stars. Nevertheless, since these measurements were carried out and handed down for hundreds of years, periodicities in the recurrence of the phenomena were identified and used for reliable predictions, the inaccuracies of the individual observations being balanced out in the long periods of time involved. Thus Babylonian astronomy, which achieved great renown, was consequently based on very simple practical fundamentals.

With respect to historical measuring instruments, the first one that may be mentioned here is the armillary sphere used by Greek astronomers. This was an ingenious instrument consisting of rings arranged one inside the other, all of which could be rotated independently. One of these carried a sighting device—notch and bead sights, as it were—which was used to take a bearing on the stars. The bearing taken could be read from the angle-divisions of the other rings. Hipparchus, who lived in the Second Century B.C. in Alexandria and later on the island of Rhodes and was probably the greatest astronomer of Classical Antiquity, used an armillary sphere about 130 B.C. to make a list of the positions of several hundred stars. This catalog provided the basis for one included by Ptolemy in his *Almagest;* it is astonishingly accurate, occupying an honored place in the history of astronomy.

By comparing new observations of the positions of stars with those of earlier astronomers, Hipparchus found that there were differences of a systematic character which he correctly interpreted as a shift in the point of reference of the measurements, the intersection of the Equator and the Ecliptic. This small movement, termed precession, is caused by the gravitational effect of the Sun, the Moon and the planets on the rotating Earth, as was realized more than 1,500 years later.

For the more accurate designation of the stars identified, this star catalog by Hipparchus also included for the first time fairly accurate information about

17 This cuneiform text from Assyria was found during excavations at Assur, the ancient capital on the Tigris. It is on a clay tablet of the Eighth or Seventh Century B.C. and contains a commentary on the fixed stars with various details about their positions, appearance and visibility.

18a, b The celestial signs in use today owe a great deal to Babylonian astronomy. The illustrations show engravings more than 2,000 years old which, from the lettering, are clearly drawings of constellations from the Late Babylonian Period. In the upper part (a), the Moon is shown between Aries and the group of the Pleiades. The lower part (b) depicts the constellations Leo and Hydra with the bright planet Jupiter at the left-hand edge of the picture.

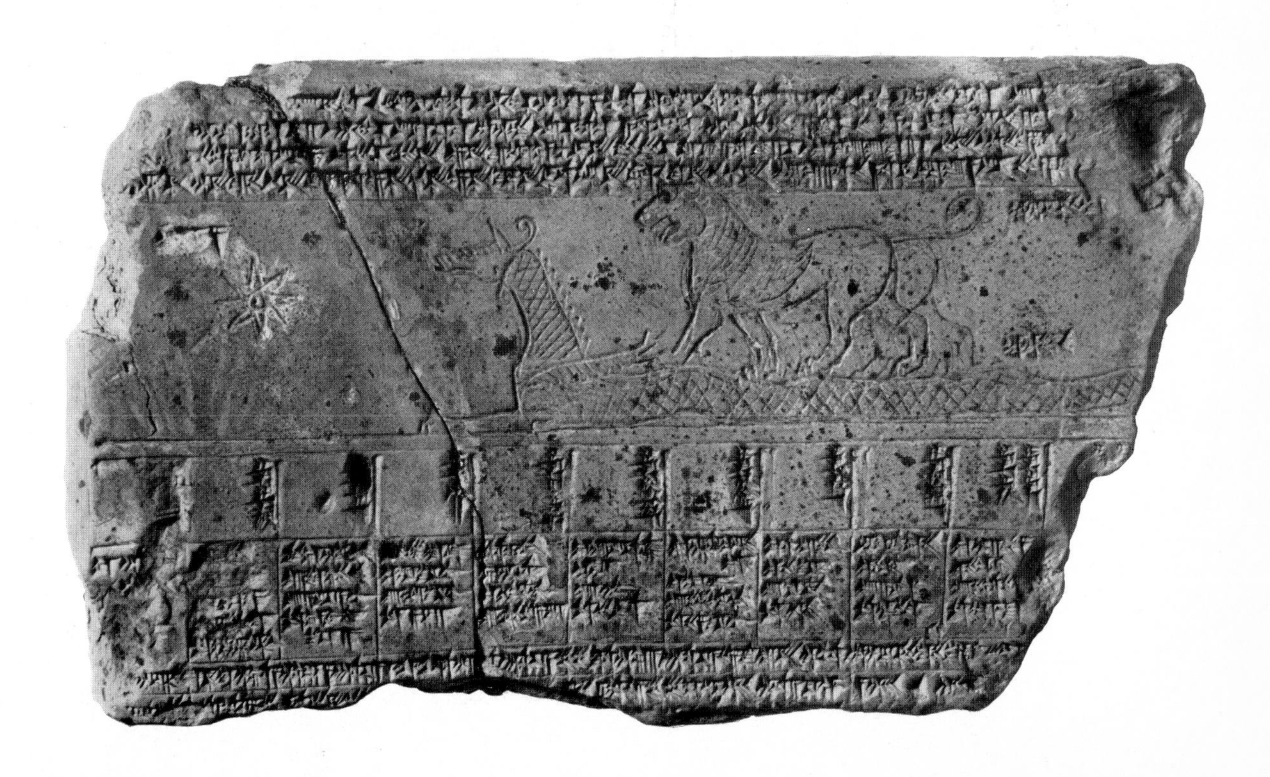

the brightness of the stars. According to their degree of brightness, stars were numbered from *one* to *six*, *one* designating the brightest and *six* the dimmest stars which could be discerned with the naked eye. How Hipparchus arrived at this division is not known, but in principle the same system is still in use today to indicate the brightness of the stars. In this connection, we still speak of "magnitude classes" or "star magnitudes," but our scale of brightness is based on objective measurements and has been extended to include the faintest stars which can only be seen with large telescopes.

Altitude-measuring quadrants

With the armillary sphere, the bearings were read off full circles but these were reduced to quarter circles when quadrants came into use. Ptolemy (about 90—160 A.D.) describes an instrument of this kind which, however, was mounted on a north-south wall and consequently only permitted altitude measurements along the meridian. The idea of a quadrant which could be moved in any direction soon suggested itself, and very many of these instruments were used as meridian quadrants and movable quadrants by later observers. In the course of time, the dimensions of these quadrants became greater and greater, the aim being to achieve a fine enough division of the scales and thus great reading accuracy despite the limited degree of precision possible in the making of the instruments.

This development was initiated by the Arabs. Continuing the creative pioneering work of the Greeks, they raised astronomy to a new and splendid level in the Eighth Century which, with the spread of Islam, ultimately influenced wide areas of the Old World. One of the principal factors motivating their endeavors to build large instruments was the desire of Ibn Carfa for a circle bounded by the pyramids on one side and the Mocattam Mountains, situated east of Cairo, on the other. Even though this was pure fantasy, the historical sources do mention mighty quadrants at the observatories in Bagdad, Damascus, Cairo, Spain, North Africa, the former Persian Empire and Samarkand. Reference may also be made to the observatories of the Indian ruler Jai Singh II, which were erected in the first half of the Eighteenth Century. The great stone instruments are evidence of a profound interest in astronomy. They were, however, an anachronism since they were built in the epoch after Tycho Brahe, Galileo and Newton.

The ingenious astrolabe

Also worth mentioning is the astrolabe, a very handy instrument which remained in service for hundreds of years. It was certainly used by the Arabs and may even date back to Hipparchus. It is employed for measuring altitudes, but its importance is due to the combination of the measuring instrument with a kind of calculator. Its ingenious design enables a whole series of astronomical problems of time and orientation determination to be solved by simply reading the answer from the instrument, without the need for tedious numerical calculations. This "calculating" function is based on the representation of certain lines originating on the celestial sphere, such as the horizon and the Equator, above which a projection of the starry sky is mounted in the form of a movable plate pierced by a number of holes. With the help of altitude measurements made with the sighting part of the instrument, the "sky" of the astrolabe can then be brought immediately into the correct position and can provide the solution of the problems in question. The design and construction of this instrument reveal an understanding of the happenings in the sky and a sound knowledge of mathematics. The same principle is still employed in the pivoted star maps or planispheres of the present time.

Tycho Brahe

If, starting with the Arabs and going from Spain to Central Europe, the history of astronomical measurements is examined, the next outstanding figure who attracts our attention is Tycho Brahe. He was born in 1546 in Scania, at that time a Danish province of Scandinavia, and his name is associated with a reform of practical astronomy. While he was still very young, he was one of the first to appreciate the need for really systematic observations to achieve further progress in this science and in particular to decide between the merits of the two rival planetary theories, the Ptolemaic system on the one hand and the Copernican on the other. Information about the positions of the planets, calculated on the basis of the two concepts, was available in the form of the "Alphonsine Tables" and the "Tabulae Prutenicae."

Nowadays, it seems obvious that the question could be decided by measurements and comparisons with the aid of these tables, but at that time this meant breaking away from scholastic habits of thinking.

19 Our knowledge of the astronomy of the Mayas originates in part from the *Codex Dresdensis*, a manuscript preserved in the Saxon Regional library of Dresden. This work is of a very late date, no earlier than the Twelfth or Thirteenth Century, and is probably a "new edition" of older originals. It comprises a series of astronomical tables on various subjects. The "Venus Table" contains data on the appearance of Venus as a morning and evening star for more than three centuries. It shows that this celestial body was of major importance for the Mayas.
Apart from astronomical data, the "Venus Table" also has pictures of deities. In this case, it is Venus as the Morning Star Regent who occupies the Celestial Throne. The attitude of the figures in the lower pictures and the coloring symbolize positive or negative aspects of the influence of the gods on life on Earth. Red is associated with a favorable disposition while blue-grey is an indication of adversity.

20 In the Maya manuscripts there is also a depiction of the "end of the world." Three deities participate in the "destruction of the world." The Celestial Dragon floods the Earth, torrents of water gushing from its throat. A "black god" attacks the Earth with a spear and other weapons. The hostile character of a female figure is characterized by a headdress of snakes and crossed bones on her clothing. The three deities destroy the world in a great cosmic catastrophe.

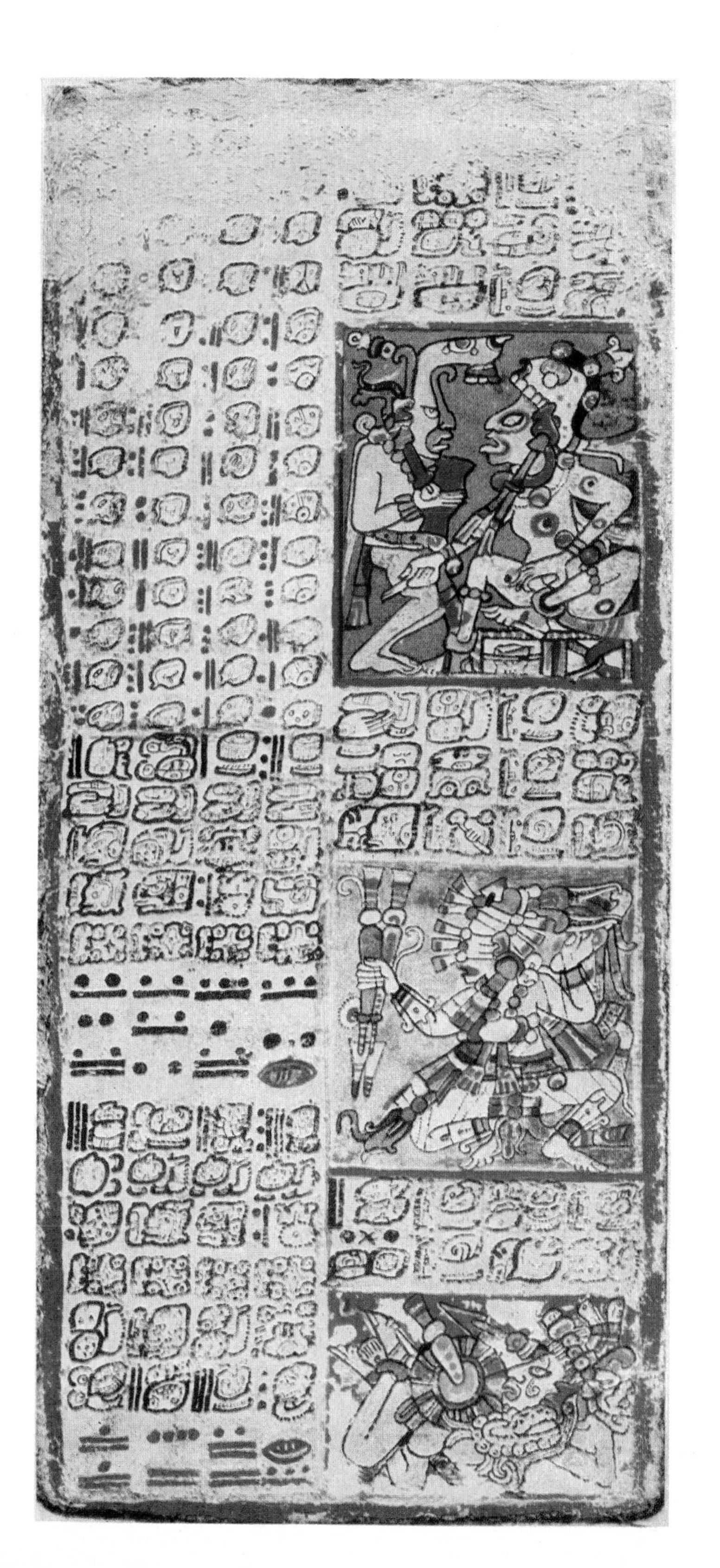

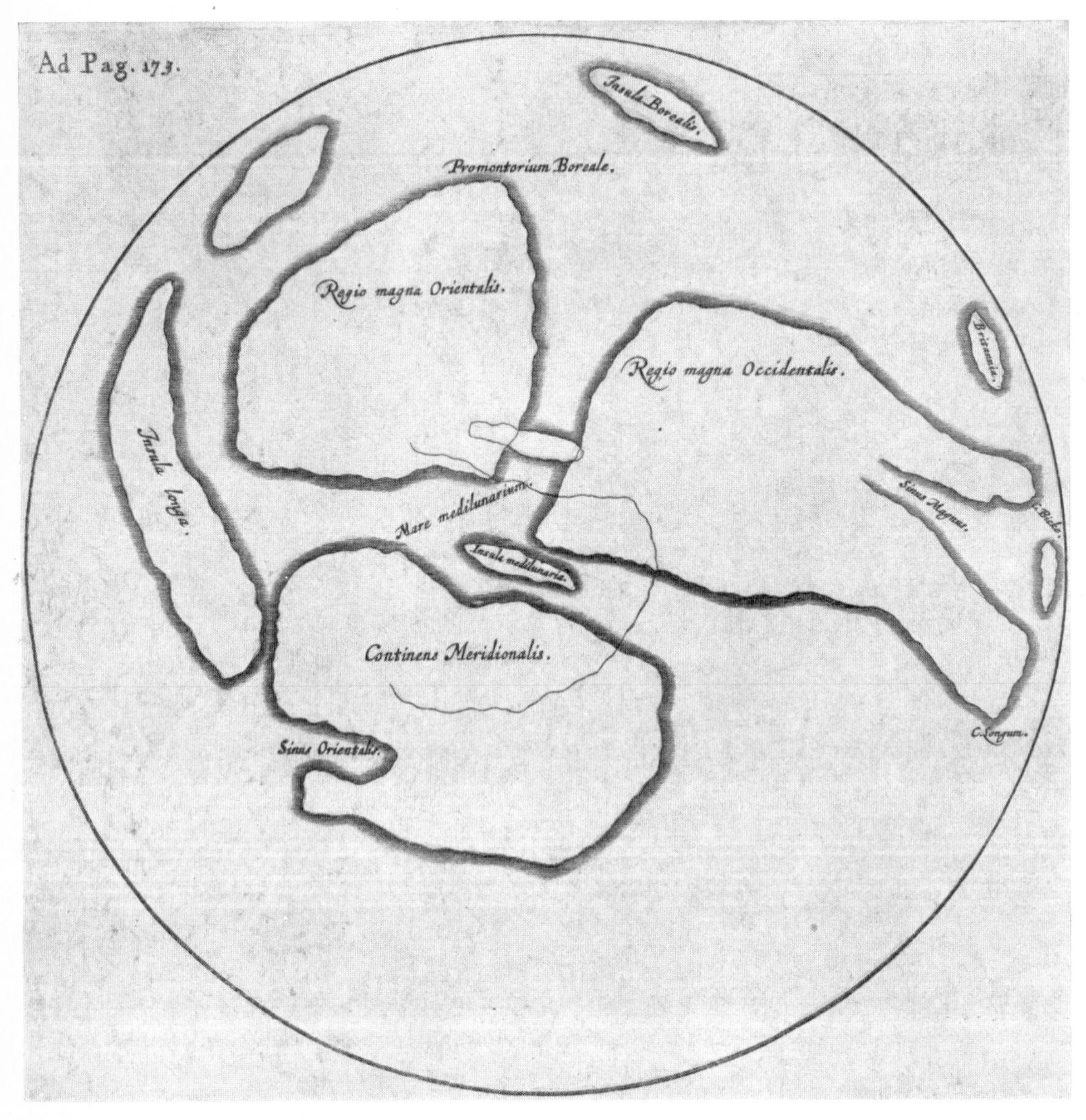

21 The only significant map of the Moon which was drawn before the age of the telescope was prepared by Gilbert who achieved fame for his discovery of terrestrial magnetism and died a few years before the invention of the telescope. From the viewpoint of the history of science it is an interesting fact that he drew his map with the intention of providing a comparative basis for identifying possible long-term changes on the Moon. For the end of the Sixteenth Century, this was progressive thinking of an extremely high order.

It may be noted at this point that Tycho Brahe subsequently elaborated his own theory about the layout of the planetary system. This stated that the planets did indeed move around the Sun, as in the Copernican theory, but the Sun, together with the planets, moved around the Earth. By this, he avoided certain difficulties which were supposed to result from the heliocentric system in practical applications and also circumvented the theological problem of the time represented by a moving Earth. Tycho Brahe had charged Johannes Kepler, his assistant in his last year of life, with the task of finding the proof of this theory and urged him to continue his work in this direction. Brahe's observations of the planets enabled Kepler to realize that all the planets, including the Earth, moved around the Sun and to understand how this was so. This marked the completion of what may be termed the "Copernican revolution," and the basic Copernican concept of a world ordered in as simple a manner as possible was proved correct.

22 This illustration gives an impression of Galileo's study from a reconstruction in the Deutsches Museum in Munich.

In addition to his consistent advocacy of the heliocentric system, the Italian scientist is equally famous for his achievements in the field of physics.
The long inclined plane shown in the picture recalls his investigation of the laws of free fall of bodies by rolling balls down such a piece of apparatus.

23 Among the ruins of the city of Chichen Itza in the north of the American peninsula now known as Yucatan, there is an ancient Maya astronomical observatory believed to date from the Tenth Century. Through the windows in the upper part, observations of the stars were made which were subsequently incorporated in the highly specialized astronomical knowledge of this civilization.

24 The basic design of the armillary sphere has remained unchanged throughout the centuries. The instrument shown dates from the second half of the Sixteenth Century. The positions of individual bright stars in particular are marked by arrows of metal. From the setting of the height of the poles it is believed that it was used by Tycho Brahe.

But, to return to Tycho Brahe, this astronomer attained his greatest fame on Hven in Öre Sound. He had been enfeoffed with this island by his patron, King Frederick II of Denmark, and it was there that he erected the observatories of Uraniborg and, later, Stjerneborg. For more than twenty years, an observatory was operated here under his direction, equipped with numerous instruments and staffed by many assistants. Close contact was maintained with scholars in many countries. By improvements of vital importance to scales and diopters, which are the measuring parts of the instruments, and through great care in their manufacture, Brahe achieved a substantial increase in the accuracy of his measurements during his stay on Hven. He did not simply accept the errors unavoidable in a measuring instrument, but corrected them subsequently in the results obtained with the aid of special tables, just as he had done as an eighteen-year-old student in Leipzig. The great achievements of this man ended when he left the island of Hven in 1597. After an unsettled period, the last years being spent in Bohemia as the astronomer of Rudolf II, he died in Prague in 1601. Of the records of his observations that he left behind, Kepler said that they were worth keeping with the priceless princely jewels. The ill will and hatred which drove Tycho Brahe from Denmark were the reason why the observatories on Hven were allowed to fall into a state of ruin. Fifty years after his death, very little remained of those places where pre-telescopic astronomy had attained its highest and final state of development.

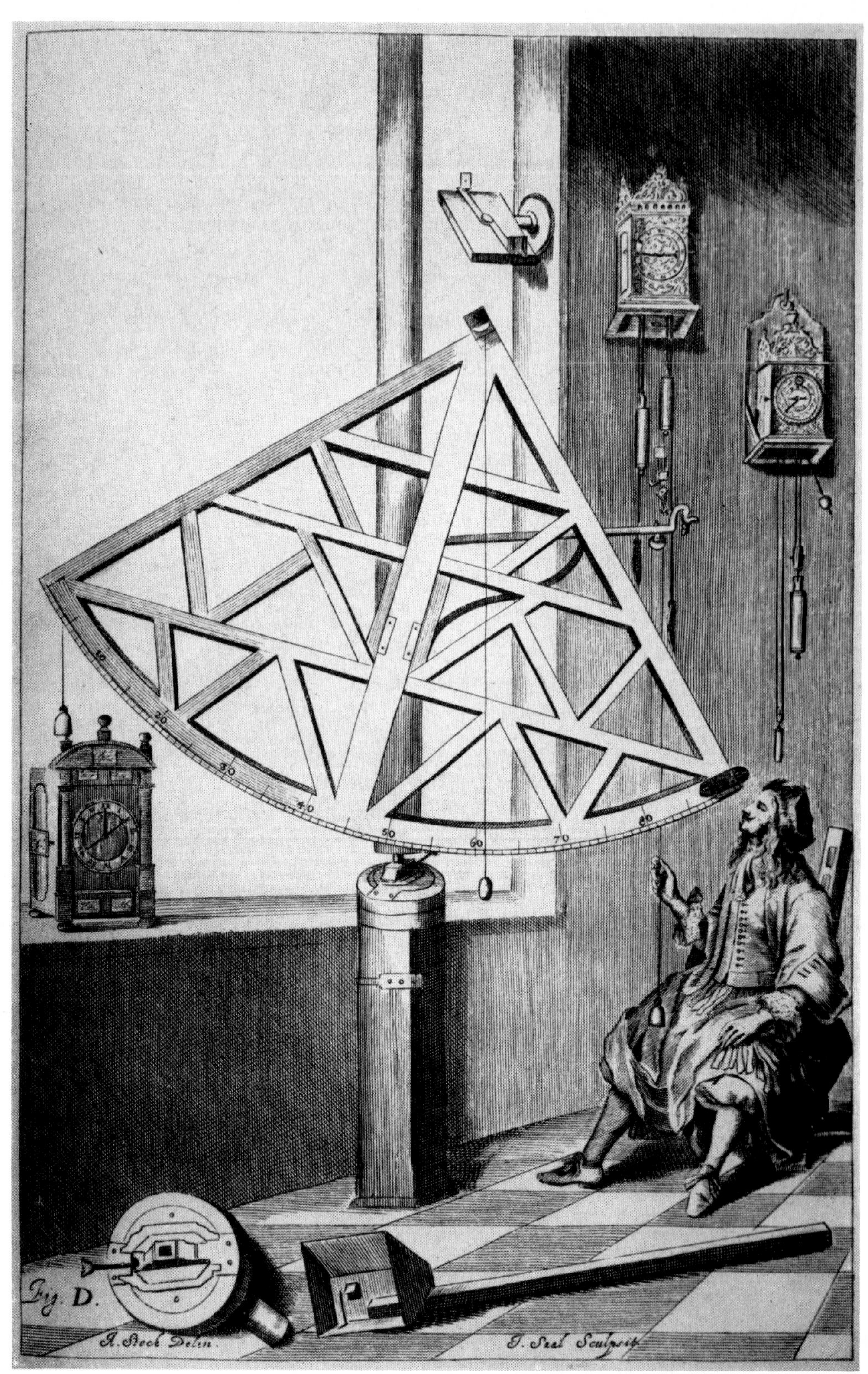

Galileo and the astronomical telescope

A great invention—the astronomical telescope—suddenly widened human horizons and provided man with a completely new opportunity to examine the wide world around him. The melting and processing of glass as the technical raw material for the making of telescopes had been known for a very long time. Ground lenses of rock crystal were found in a stratum at Troy dating from 2000 B.C. These may well have been used primarily for decorative purposes but their magnifying effect was certainly known. It is definitely known that spectacles appeared in Europe only in the Middle Ages. There was occasional speculation about glasses with which it was hoped to see distant objects more clearly, as by Roger Bacon in the Thirteenth Century and Leonardo da Vinci two hundred years later, but it seems likely that the first

25 This azimuth quadrant was used in the Seventeenth Century for measuring the altitude of stars. The instrument had a horizontal pivot mounting and a vertical column. By adjusting the two mountings, a sighting could be taken of any star with the diopter. The altitude could then be read from the point on the scale indicated by the plumb line.

26 In the city of Jaipur, 200 kilometers (124 miles) to the southwest of New Delhi, there is one of the five observatories built by the Indian ruler Jai Singh II. It was completed in 1734. The staircase in the center of the picture forms the center of a great sundial. The position of the Sun was measured in the polished bowls in the foreground.

telescope was not constructed before 1600. Several people have a claim to have been the first with this invention, but it seems that Johannes Lippershey of Middelburg in the Netherlands may justly be regarded as the original inventor.
His interest awakened by reports about the magnifying effect of suitable combinations of lenses, Galileo Galilei (1564—1642) also started experimenting and by 1609 had made an instrument of approximately three times magnification. Telescopes of eight and even thirty times magnifying power rapidly followed, the latter with a lens diameter of about 5 centimeters and a length of 1.24 meters. It may also be noted that the word "telescope" was coined by the group headed by Galileo.
As at other points in the history of astronomy, in this connection, too, the name of Galileo must be coupled with that of Kepler (1571—1630). Galileo made the actual telescope, but it was Kepler, with a more profound understanding of the optical qualities of lenses, who supplied the theory for it. It is to be found in his book *Dioptrice*, written in Prague. Kepler expressed his appreciation of the new "wonder instrument" in the words, "Oh, thou tube of knowledge, more precious than any sceptre."[1] Even though the most prominent of his achievements is the explanation of the true movement of the planets, his contribution to geometrical optics should not be forgotten.
In general, Galileo is considered to have been the first to use the telescope for astronomical observations. Indeed, it cannot be asserted that none of his predecessors in the construction of these instruments had ever taken a look at the heavens, but Galileo was the first to describe phenomena which had hitherto remained unseen and to draw very far-reaching conclusions from them. Above all, he immediately recognized the significance of the new findings for demonstrating the validity of the Copernican theory of the planets.

Making Moon maps and observing fixed stars
At the same time as Galileo's Moon maps, drawings of our satellite were produced in England which were clearly prepared with the aid of a telescope and were more accurate than those of Galileo. The spread of the telescope throughout Europe at this

27 The monk in the center of the group is using an astrolabe for measuring the altitude of a star. He is taking a bearing on the object through the diopter-sight pivoted at the center of the disk.

28 This brass astrolabe dates from the middle of the Sixteenth Century and was made by Johannes Praetorius of Nuremberg. It is 39 centimeters (15 inches) in diameter. The "mater" of the astrolabe carries the projection of selected imaginary lines of the sky, e.g., the numerous lines of equal height above the horizon. Over this, there is a pierced plate, known as the "rete," the points of which mark the positions of the brightest stars. By turning the mater and rete in relation to each other, the relations between the stars and the lines can be reproduced, and the corresponding time can be read off the outer scale which is divided into hours. On the other side (not shown), there is the scale needed for the measurement of the angle of altitude and the diopter-sight.

29 Model of the Uraniborg, based on Tycho Brahe's description of his observatory. The building, seen here from the southeast corner, contained the living quarters of the Brahe family, studies, guest rooms, and even a chemistry laboratory. Two instruments for altitude measurement and an armillary sphere can be seen on the observation platforms and in the garden.

30 In the middle of the Sixteenth Century, an observatory of great significance in the history of astronomy was founded in Kassel by the Landgrave Wilhelm IV, a passionate astronomer and keen observer. Before building the Uraniborg, Tycho Brahe paid a visit here, found stimulation for his work and kept in touch with the Landgrave until the latter's death. The "Wilhemsquadrant," a brass instrument about a meter (39.37 inches) in height, was an example to Tycho Brahe when he built his own instruments. A feature of special interest for that time are the four screws for the accurate horizontal adjustment of the base circle.

time was in any case exceptionally rapid. This was due to the upswing in technology and science which had taken place in the Renaissance and during the period following it as a consequence of the rapid development of trade and commerce at the end of the Middle Ages. The new social forces which were then emerging represented an economic stimulus which in turn was the condition for technical and scientific progress.

Initially, the telescope was primarily employed for descriptive observations and, to begin with, was seldom used for determining positions. The attitude of Johannes Hevel (Hevelius), an astronomer in Danzig, is characteristic of this. He made a careful and systematic study of all celestial phenomena and used the telescope to observe eclipses, the Sun, the Moon, the planets and so on. Oddly enough, however, Hevelius was totally opposed to the use of the new instrument for measuring positions. Nevertheless, even without the assistance of a telescope, he achieved such an incredible degree of accuracy that Edmund Halley was despatched to Danzig in 1679 by the Royal Society to ascertain whether Hevelius' findings were genuine. Halley had to give Hevelius a splendid testimonial since he himself, with the observations he made with a telescope, could not attain a higher degree of accuracy than the latter with his traditional "sighting" methods.

It is clear that this attitude to the use of the telescope could not be maintained for very long. The naked eye was good enough for describing movements within the planetary system but in the final analysis the greater accuracy of telescopic observations was essential for proof of reciprocal changes in the positions of the fixed stars. A proof such as this was becoming increasingly more urgent. One of the conclusions to be drawn from Copernicus' theory was that the movement of the Earth around the Sun had to be reflected in the periodical change of direction of the fixed stars. Parallactic motion is the astronomical term for this. Many attempts had been made to identify this phenomenon. Even Copernicus had tried to obtain proof of the validity of his model in this way, but the technical means at the disposal of astronomers had hitherto been insufficient.

An initial success was achieved by James Bradley (1692—1762) who carried out his work at Oxford and at Kew near London. However, as often happens in science, it was something else that he found. The position of the star Gamma in the constellation of Draco had been accurately measured with a special instrument. It was Bradley's intention to repeat this measurement six months later. By then, the Earth, and with it the observer, would have moved far away from its first position, and it should have been possible to identify a small change in the position of the star against its background. Nothing could really be expected by another observation only a few days after the first, but Bradley's curiosity was so great, as he himself remarked in a subsequent letter to Halley, that he again took the bearing of the star. There was a slight change in its position! Puzzled by the reason for this, Bradley carried out further checks and eventually was sure that he had made an unexpected discovery. Just as drops of rain meet an oncoming car from the front and run down the windows at the side in a diagonal manner, when light meets the moving Earth it is somewhat deflected in the "direction of travel." In principle, this finding was enough to prove Copernicus' concept: the Earth was following some kind of orbit. That it had still not been possible to identify the expected parallactic motion only indicated how far away the stars must be. On the basis of the accuracy of measurement achieved, Bradley estimated that his star in Draco is more than 400,000 times as far from us as the Sun. He thus inferred a correct order of magnitude for the tremendously great distances of the fixed stars.

Discovery of distance

The measurement of the parallaxes of fixed stars was ultimately achieved in the first half of the last century. Friedrich Wilhelm Bessel (1784—1846) found the long-sought change in position after a series of carefully planned observations of star 61 Cygni. On its annual journey, this little star moves away from the central position by only 0.3″, a tiny angle indeed. The entire movement could be completely concealed by a coin of less than two centimeters in diameter held at a distance of ten kilometers! This was the first real step forward regarding measurement in the area beyond the planetary system. The method employed was of fundamental importance. When distances are nowadays stated for the most remote stellar systems, they ultimately rest on fixed star parallaxes of this kind. The distance to a stellar system cannot be directly measured. To learn something about this distance, use has to be made of the "special characteristics" of the object. These characteristics require calibration, however, and with the

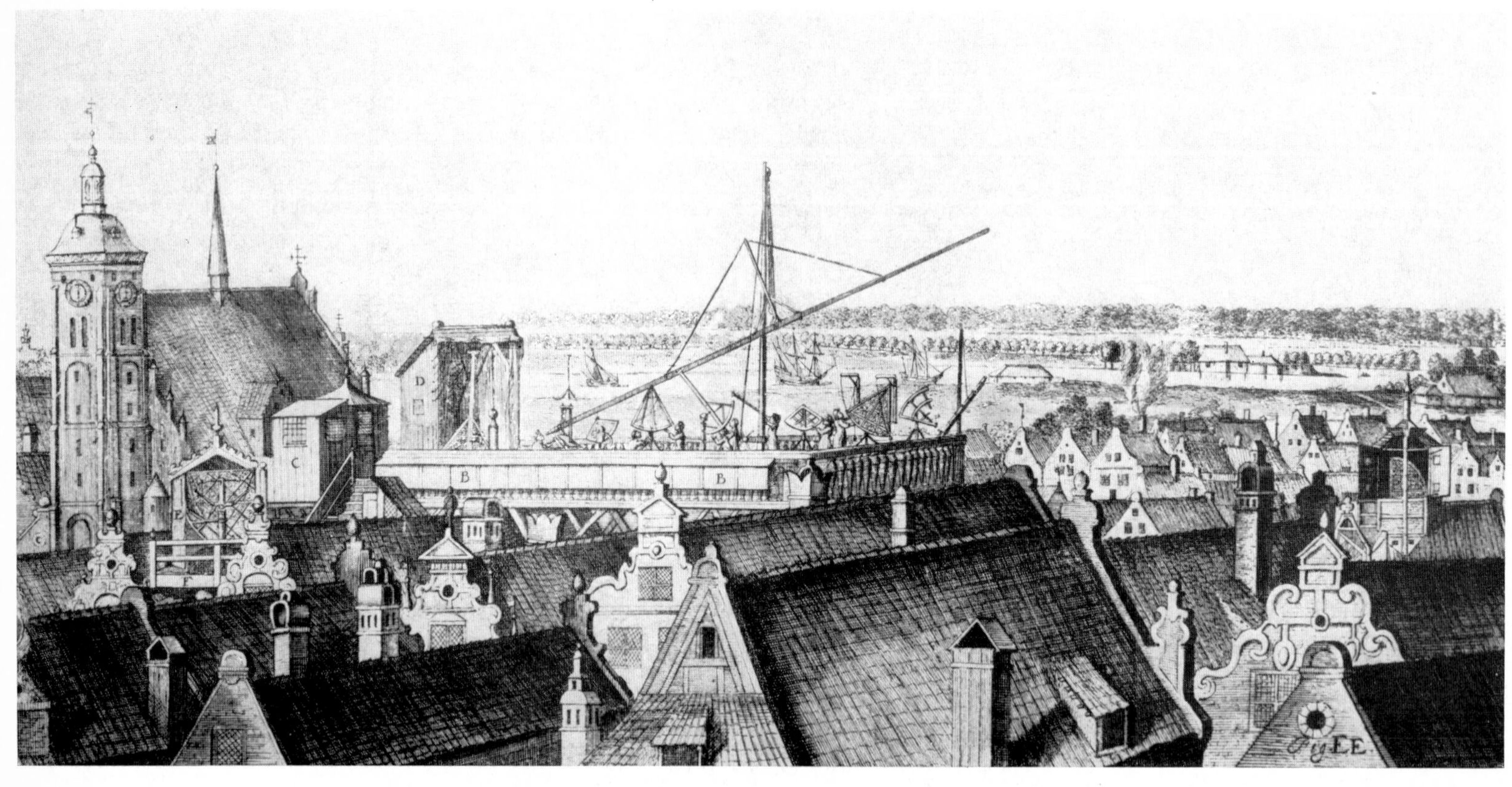

gradual inclusion of celestial bodies closer to the Earth this calibration ultimately ends at those few stars which, in the course of the year, allow the deviation in question to be identified. Distances are of basic importance when it is a matter of making precise statements about the physical processes in celestial bodies. This is why there is some justification for saying that Bessel helped to give us access to astrophysics, although this was not at all his intention. His contribution to astronomy is of inestimable value. In the memorial address read on the occasion of his death, the speaker raised the question of what was most admirable about him: "... the number and excellence of his theoretical papers, the acuteness of his observations or their profusion."[2] And yet Bessel had declared that the function of astronomy was simply to seek the laws which would enable the motions of the celestial bodies to be predicted as accurately as required. Anything else which might otherwise be observed about the stars—perhaps this is formulated in an exaggerated manner—was not really of any interest to astronomers! This restriction, expressed by such an authority, exerted a major influence on the development of astronomy for a long time. Only a century ago and less, the principal instruments of European observatories were employed to determine the positions of the stars as the basis for deriving their movements.

Telescopes: refractors versus reflectors

In retrospect, the technical history of the telescope can be seen as a contest between the refracting telescope and the reflector. The development of the telescope started with the refractor (another term for the lens telescope). Glass lenses "bend" the rays of light coming from the object and striking the eye so that things appear at a greater angle, i.e., nearer. The same optical effect can be obtained by substituting a concave mirror for the objective lens. In actual fact, the refractor was soon followed by the reflecting telescope. Both types of instrument had their characteristic shortcomings and, as these were gradually overcome in the course of time, each kind became more or less popular in turn. The lens telescope, fitted with a single objective lens to begin with, suffered from poor image-quality, due to colored blurs

31 The illustration shows the observatory founded by Hevelius in 1611. Johannes Hevelius was the son of a rich merchant of Danzig and became a well-known astronomer after his retirement from business life. The most striking instrument is the telescope with its long focal length. After being the scene of very intensive work for nearly forty years, the observatory was burnt down by an arsonist.

32 This interior view of Johannes Hevelius' observatory shows the image of the Sun projected by a telescope on to a screen. The measuring instruments hanging in the background were used in-depently of telescopes but were typical of Hevelius' observation technique in determining positions. The engraving is taken from *Machina Coelestis*, one of the works of Hevelius, in which he describes his instruments, his observations being analyzed in a second part.

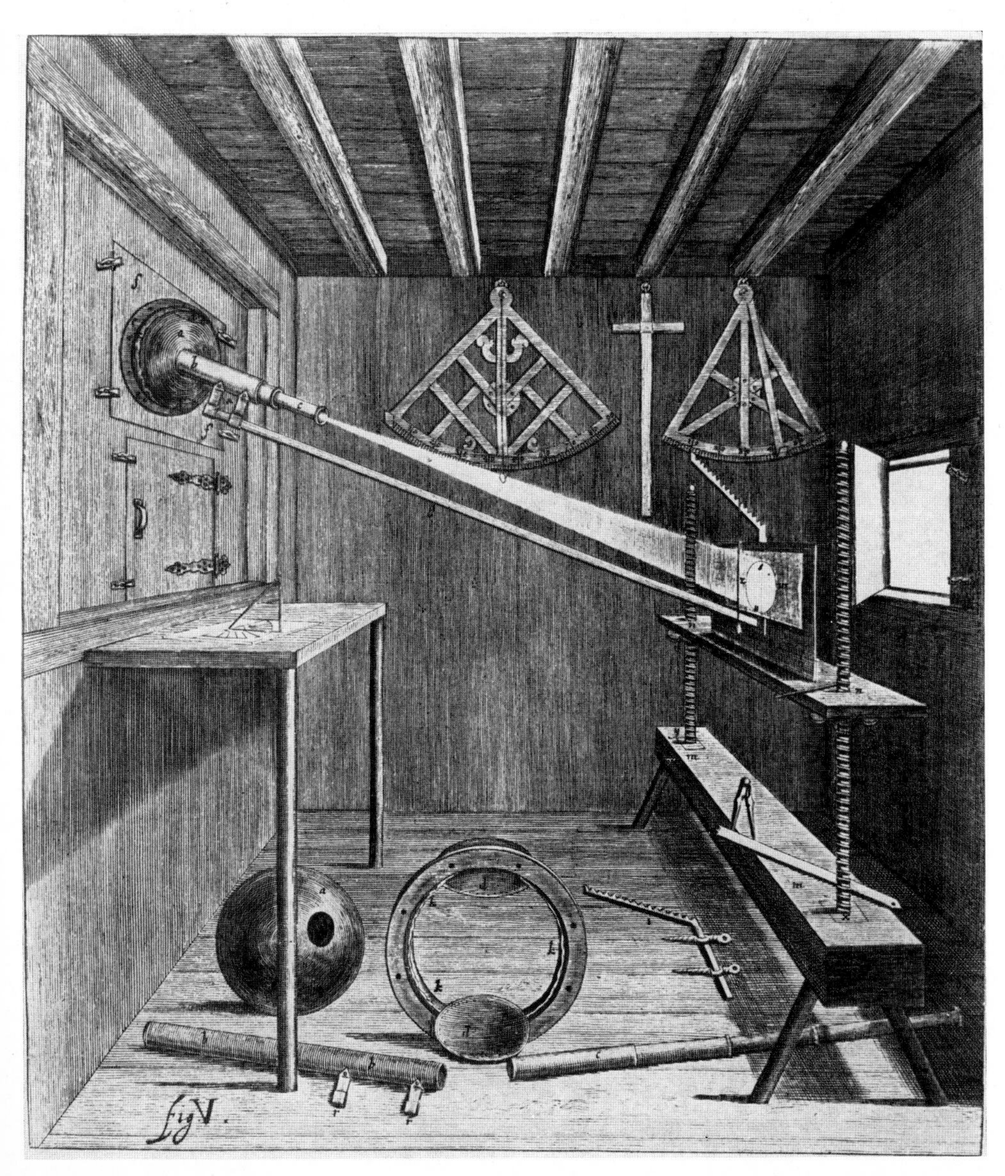

around the edges of the object observed. The explanation of this is that the single lens has a varying effect on the components of different wavelength contained in the incoming light. This results in images of different colors which do not completely coincide. This phenomenon is known as chromatic aberration. Mirrors are free from color faults of this kind, but the problem here was to find a suitable material for their manufacture. They were made from metal or metal alloys, but their reflective qualities were far from ideal and the effect of air on them also caused rapid deterioration. Then there were difficulties in the accurate grinding of the surface. Hoping that improvements would prove possible later, even Sir Isaac Newton (1643—1727) broke off his experiments with reflectors of his own manufacture. In actual fact and within his own lifetime, a countryman of his showed the Royal Society a reflecting telescope which was considered excellent by such an authority as Bradley, among others. The predominance achieved by reflectors was not initially of long duration. Around the middle of the Eighteenth Century, refractors with achromatic compound lenses were built in England by John Dollond and others. It is an interesting fact that the considerations which preceded this development were based on the human eye, which was known to be composed of two different media. Since it was also assumed—incorrectly—that the eye is free from color faults, it seemed obvious that achromatic lenses could be obtained by the combination of different kinds of glass. This was indeed the case, despite the faulty reasoning on which it was based. A period of stagnation interrupted further developments in this direction, however, since the lens makers could not cope with the problems they now faced. For an achromatic lens system, matched sorts of glass had to be selected on the basis of their optical characteristics. This meant that the lens maker had to have a sufficient number available and be capable of assessing their optical data. Furthermore, it was now necessary to grind the surface of the lenses to a specific curvature. High-grade glass with suitable optical qualities and accurate methods of measurement were needed for further progress.

Advances in optics technology

At this point, the history of optics is inseparably linked with the name of Joseph Fraunhofer. Born near Munich in 1787, he completed an apprenticeship as a mirror-maker and glass-grinder and at the same

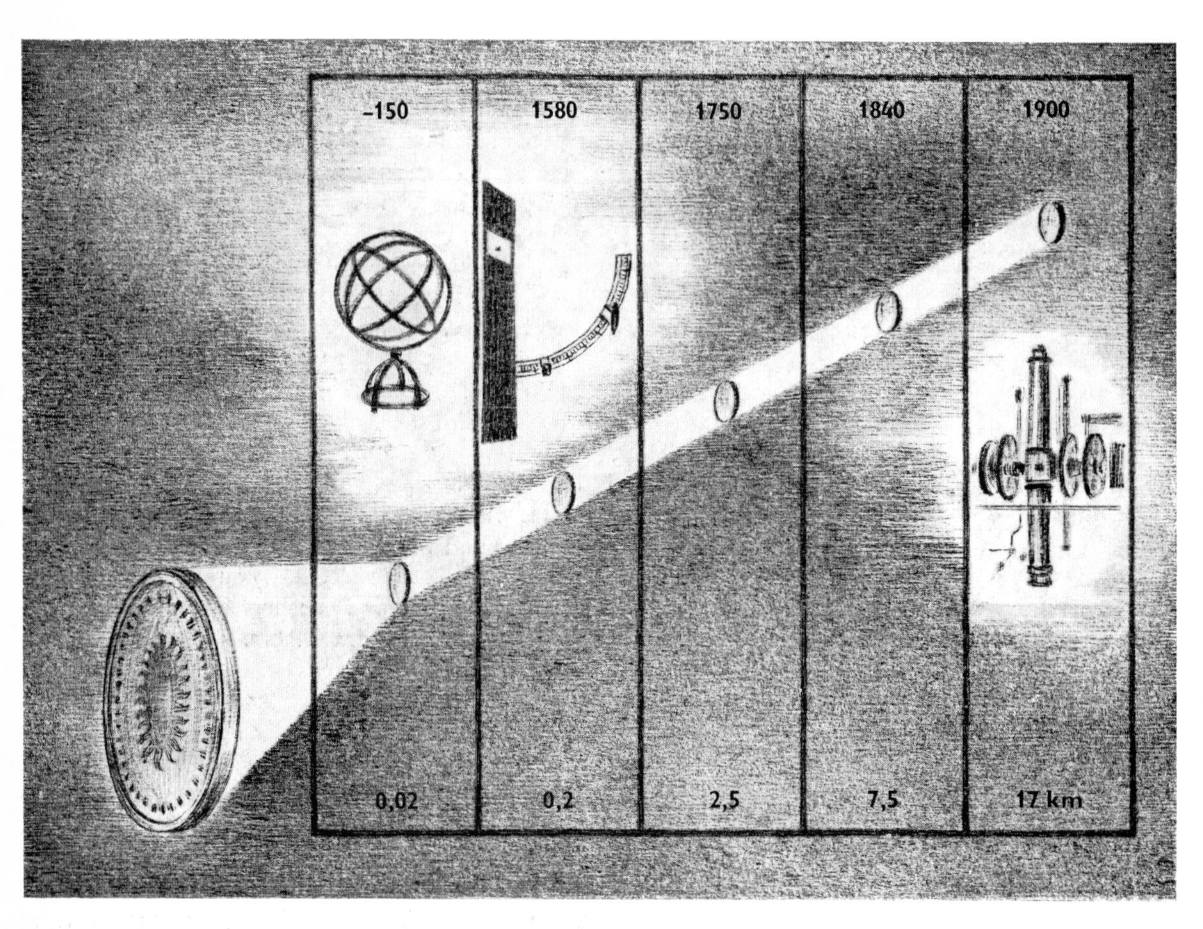

33 In the course of the centuries, considerable advances were achieved in the accuracy of position measurements. The margin of error in measurements is compared here with the apparent diameter of a coin. Thus, at the time of Hipparchus, for instance, the distance between two stars could be determined with an accuracy of four arc-minutes with the aid of an armillary sphere. This angle would be equivalent to the diameter of a coin of one inch when seen from 20 meters (about 66 feet). Through the work of Tycho Brahe (who used mural quadrants), James Bradley and Wilhelm Bessel, measurements ultimately became so accurate that by the turn of the century measurements accurate to within fractions of arc-seconds were being achieved with the aid of meridian circles. This implies a standard of accuracy equivalent to the diameter of a coin when seen from 17 kilometers (10.5 miles).

time took an interest in optics and lens-grinding. Through the collapse of a house, which he luckily survived, he attracted the attention of influential patrons. His outstanding ability was soon recognized and at the age of nineteen he joined the Munich enterprise known as the "Mathematisch-mechanische Institut Utzschneider, Reichenbach und Liebherr" as an optical craftsman. He very rapidly became the chief of the optical department of this firm which in the meantime had been moved to Benediktbeuern in Upper Bavaria. Prior to this, the far-sighted Utzschneider had established a glassworks at the same place and it was here that Pierre Louis Guinand and later Fraunhofer, too, reformed the technology of glass-making. In optics, Fraunhofer completed the step from the trial and error stage to the provision of exact process specifications based on prior investigation of the glasses to be used. Through his methods of work, not only in the grinding of lenses but also in the problems of precision engineering associated with instrument construction, the manufacture of optical instruments developed from a craft to an industry. The production program drawn up by Fraunhofer for Benediktbeuern is essentially modern in concept. He concentrated on a restricted number of types of lenses currently in demand for "quantity production" but, as a point of honor, also produced a small number of one-of-a-kind lenses of outstanding quality. The results Fraunhofer obtained were such that his optics cannot be faulted even in the light of present-day knowledge.

William Herschel's remarkable mirrors

In the year that Fraunhofer was born, when it was not yet possible to make good lenses with a diameter of more than about 15 centimeters, a reflecting telescope with a diameter of 1.22 meters (48 inches) was built at Slough, near London. It was made by William Herschel (1738—1822) who had come to England as a military musician with the Hanoverian Guards and

34 In the first half of the Eighteenth Century, the most famous maker of telescopes was James Short of Scotland. Most of his telescopes were fitted with a metal mirror pierced in the center. With this arrangement, the incoming light is directed out of the telescope through the main mirror with the aid of an auxiliary mirror. This system is still used with modern telescopes.

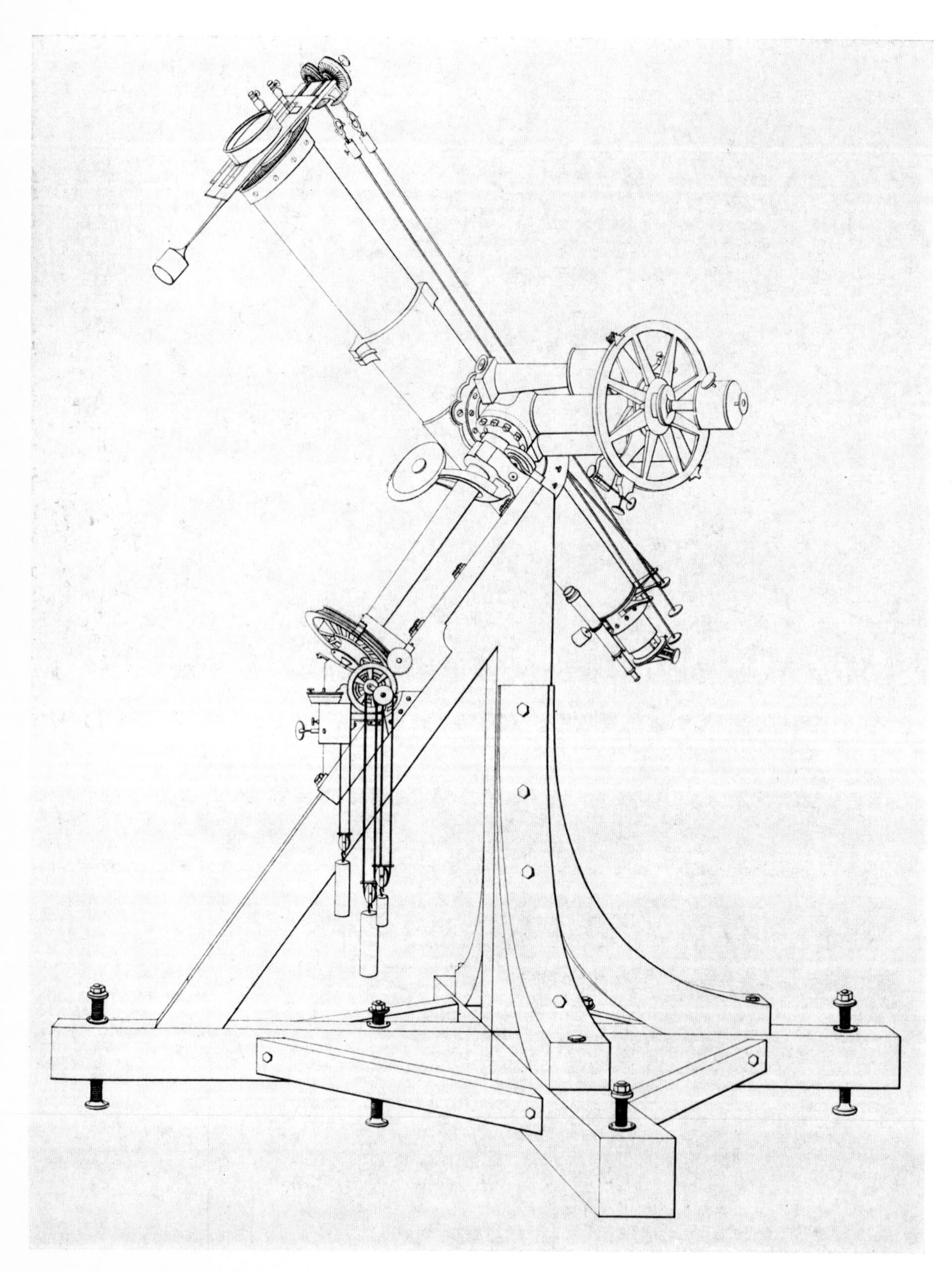

35 Using an heliometer, Bessel was able to measure the parallax of star 61 in the constellation Cygnus. This type of instrument has a divided lens, each half projecting an image of the star area in question. By a mechanism of threaded spindles, the two parts of the lens can be moved in relation to each other in a measured manner. In this way, two different stars can be made to coincide, and the angle between them can be found from the movement required for this. This instrument is one of the masterpieces produced by Fraunhofer's workshop. Every detail was devised and designed by Fraunhofer, but the actual instrument was only completed after his death.

36 The glassworks founded by Joseph Von Utzschneider was accommodated in the former washhouse of the Benedictine monastery at Benediktbeuern. Major advances in the technology of glass smelting were made here. The building is now a memorial to Joseph Fraunhofer.

subsequently settled there. He could only pursue his passionate interest in astronomy by making his own telescope since, to begin with, he could not afford the small reflecting telescopes, let alone Dollond's achromatic instruments. Herschel consequently concerned himself with the casting of metal alloys and, as he himself said, within a few years he had ground more than 400 mirrors. He selected the best of these, continued to work on them and ultimately surpassed the professional mirror-makers with the quality of his products. For him, this work was not an end in itself, and his real aim was always astronomical observation. He became famous through his discovery of the planet Uranus. As a result, he was made the King's private astronomer and achieved financial independence. Nevertheless, even at this time, he still sold successful instruments and his price for a two-meter reflecting telescope of 16 centimeter diameter was 100 guineas. A maxim, which is still valid today, is that the practicing astronomer is never satisfied with his instrument and always demands bigger and better telescopes. This was also true of Herschel. He made the observations for which he became famous with a fine instrument that had a mirror diameter of almost 50 centimeters. He finally achieved a diameter of 1.22 meters in the instrument to which reference has already been made, but it seems that he was never really satisfied with it.

In any event, it was William Herschel who introduced the age of the giant telescopes. With such an instrument, the light coming from the celestial bodies is collected on a large area, the mirror aperture and, as if through a cone, arrives at the eye of the observer. In this way it is possible to see far into Space and identify objects that are very dim. Despite certain shortcomings, the future of reflectors was now assured and their position was consolidated still further when the silver-on-glass technique for

37 There were two smelting furnaces in the glassworks from which Joseph Fraunhofer obtained his glass. They were fired with wood, their special design permitting the temperature of 1300 °C required. 200 kilograms (440 pounds) of glass was produced by each charge.

38 Joseph Fraunhofer was the first to use the equatorial or parallactic telescope mounting for fairly large instruments, too. This refractor from the Optical Institute at Benediktbeuern clearly demonstrates the principle of the mounting. The telescope tube is movably mounted on two axes, the one inclined at the column and the other horizontal and carrying the counterweight. Once a bearing has been taken on a star, the object can be kept in view by following it around the inclined axis, despite the slow daily rotation from the eastern to the western horizon.

39 This small reflecting telescope by William Herschel consists of an eight-sided wooden tube with an eyepiece at one end (upper left in picture) and the mirror at the other (lower right). A small lens telescope for seeking the stars is mounted near the eyepiece.
The mounting is characteristic of Herschel's method of construction. The tube can be brought to different vertical positions by a pulley system. Changes parallel to the horizon are carried out by turning the entire frame. A gear rack arrangement is provided for fine adjustments.

40 Completed in 1787, this reflecting telescope by William Herschel remained the largest in the world for half a century. The tube was 14 meters (46 feet) in length. For the operation of this cumbersome monster two persons were required with whom the observer, standing at the front end of the tube, communicated via a speaking tube. After a family celebration held in the telescope tube, Sir John Herschel, the son of the constructor, had it closed down in 1839.

making mirrors was introduced. It was then possible to use glass, which was superior to metal alloys, as a material for mirrors. However, refractors still continued in use and in the last two decades of the previous century excellent instruments of great size were made. The greatest of these is the 40-inch (1.02 meters) refractor of Yerkes Observatory, University of Chicago, which is 20 meters long and was used for the first time in 1897.
It has not proved possible to build larger instruments of this type since a glass disk of the size required tends to distort under its own weight, making it unsuitable for use as a lens in a telescope. The mirror of a reflector, on the other hand, is supported on the reverse side so that it is more rigidly constructed right from the start.

Refractors, reflectors and astronomical photography

Lens telescopes still remained important. Unlike the standard reflecting telescopes, it was possible with such an instrument to obtain a sharp image of whole areas of the sky, covering several degrees, at the same time. This is why refractors were of such value, especially when photography came into use for astronomical observations. However, in this field

Fig: 1.

41 One of the great lens telescopes constructed at the turn of the century is the double refractor at Potsdam. It consists of an instrument with an aperture of 80 centimeters (32 inches) for taking photographs of the sky and a second telescope of 50 centimeters (20 inches) diameter for visual observations.

42 The movement of the air which takes place during the day in strong sunlight makes it difficult to identify very tiny details in the atmosphere of the Sun. Part of the turbulence occurs directly in the domes which normally enclose astronomical telescopes. To avoid this inconvenience, the Fraunhofer Institute in Freiburg/Breisgau commissioned a refractor for solar observations without a dome and this was then erected on the island of Capri. The 35-centimeter (10.4-inch) lens brings the light through the instrument-mounting into the interior of the building and on to the slit of a spectrograph. Apart from the usual spectrographic investigations, charts of the magnetic field of the Sun can also be prepared automatically here.

also mirror instruments ultimately left lens telescopes far behind them.
In 1931, Bernhard Schmidt, working at the observatory in Hamburg-Bergedorf, achieved a great increase in the field of view of reflectors by his brilliant invention. The Schmidt mirror system now enables photographs to be made of areas in the sky covering several degrees. The largest instrument of this kind is in the Karl Schwarzschild Observatory at Tautenburg near Jena and has an effective aperture of 1.34 meters in diameter. It can be concluded that the contest between the lens telescope and the reflector has now finally been decided in favor of the latter. Large refractors are today only used for special tasks.

Spectroscopy

To explain the development of astronomical measurement techniques since the turn of the century, it is necessary at this point to say something about the subject of research of modern astronomy. Nowadays, most attention is concentrated on astrophysics. This branch of astronomy is characterized by the fact that in addition to the laws of mechanics all the other findings of physics are being utilized to obtain information about the state of cosmic bodies, their origin and development and about their position in the Cosmos as a whole. Joseph Fraunhofer, and in a certain respect William Herschel as well, created the essential conditions for this in the technical sense. At the same time, however, their work formed the foundation for new methods of research. Thus, spec-

troscopy may be regarded as a field of study which was founded by Fraunhofer.
To be able to assess the glasses he produced, Fraunhofer had to carry out investigations at different wavelengths. This inevitably caused him to pay close attention to the spectrum of sunlight, obtained through prisms. In the course of this, he discovered the dark absorption lines now known as Fraunhofer lines. Comparison with the dispersed light of other stars showed "... that the light of Venus in this connection is of the same nature as sunlight."[3] On the other hand, he found no similarity with the spectrum of Sirius, a bright star. In principle, comparisons of this kind are still used to classify the spectra of the stars in spectral classes which are distinguished from each other by physical criteria.
It is the merit of Gustav Kirchhoff (1824—1887) and Robert Bunsen (1811—1899) that they investigated the problem of the origin of the Fraunhofer lines and raised spectroscopy to a method of research. It is related that it was a firework display in the Castle at Heidelberg that first drew their attention to the relationship between the light emitted by colored flames and the chemical composition of the light source. They developed the spectroscopic method for the chemical analysis of terrestrial substances. By extending this method, Kirchhoff in 1861 attributed the Fraunhofer lines in the spectrum of the Sun to the presence of known chemical elements in its atmosphere. He also concluded from this that the Sun must be an incandescent body. Hypotheses such as this had already been made before, but this was the first time that they followed from observations. Kirchhoff was quite aware of the significance of his work, and for him "... the more exact exploration of the spectrum of the Sun [was] of no less importance than that of the fixed stars themselves."[4] The true significance of this method only became apparent, however, when technical advances enabled it to be applied to the stars. Developments in this field were rapid when, at the beginning of this century, photography replaced the visual observation of the spectra.
Astronomers obtain spectra with the aid of spectrographs which are used as accessory instruments in association with telescopes. They are mounted on the tube of the telescope, arranged on the actual mount of the telescope so that it is moved simultaneously with it or, as in the coudé system, is located in a non-movable manner in a basement of the domed building that houses the telescope. The light of the star must then be passed to it through the axis of the telescope. In this case, spectra can be resolved into finer details and, above all, the use of larger optical components permits greater efficiency.
A wealth of information can be obtained from the evaluation of the spectrogram. The component of motion of the object along the optical path can be ascertained from the position of the absorption lines. Together with the results obtained by traditional observations, which can in some circumstances identify changes in position on the celestial sphere, a spatial picture of the motions of celestial bodies can be obtained. These observations mainly concern single stars, double stars, interstellar clouds and extragalactic objects. The width of the lines allows conclusions to be drawn about the physical conditions in these objects. In this manner, information can be gained not only about the pressure and temperature in the outer strata of the stars but also about the total amount of radiation emitted. Furthermore, quantitative assessments can be made of the abundance of chemical elements in the stellar atmospheres. Finally, the shape of the lines allows conclusions about the internal motions of the light-emitting matter and about the presence of magnetic fields.
It seems incredible that the analysis of a spectrum can supply so much information. A blackened strip only a few millimeters long on a photographic plate is frequently the reward for hours of exposure with the aid of a great telescope. Yet even such an apparently disappointing result as this can provide data on the state of motion of an object millions of light-years away and, ultimately, about the structure of our Universe. All this can be more easily appreciated perhaps when it is realized that generations of scientists, working in the various disciplines, have concerned themselves with methods for obtaining, evaluating and, above all, interpreting spectra.

The examination of Space

If it was Fraunhofer's work that prepared the way for the investigation of the individual stars, then it can be said that it was William Herschel who examined Space itself. As he himself remarked, his predecessors conceived the starry sky as the concave surface of a sphere surrounding the eye of the observer. This was adequate enough so long as the stars were only used as reference points for the determination of distances within the solar system. In actual fact, the differences in the apparent bright-

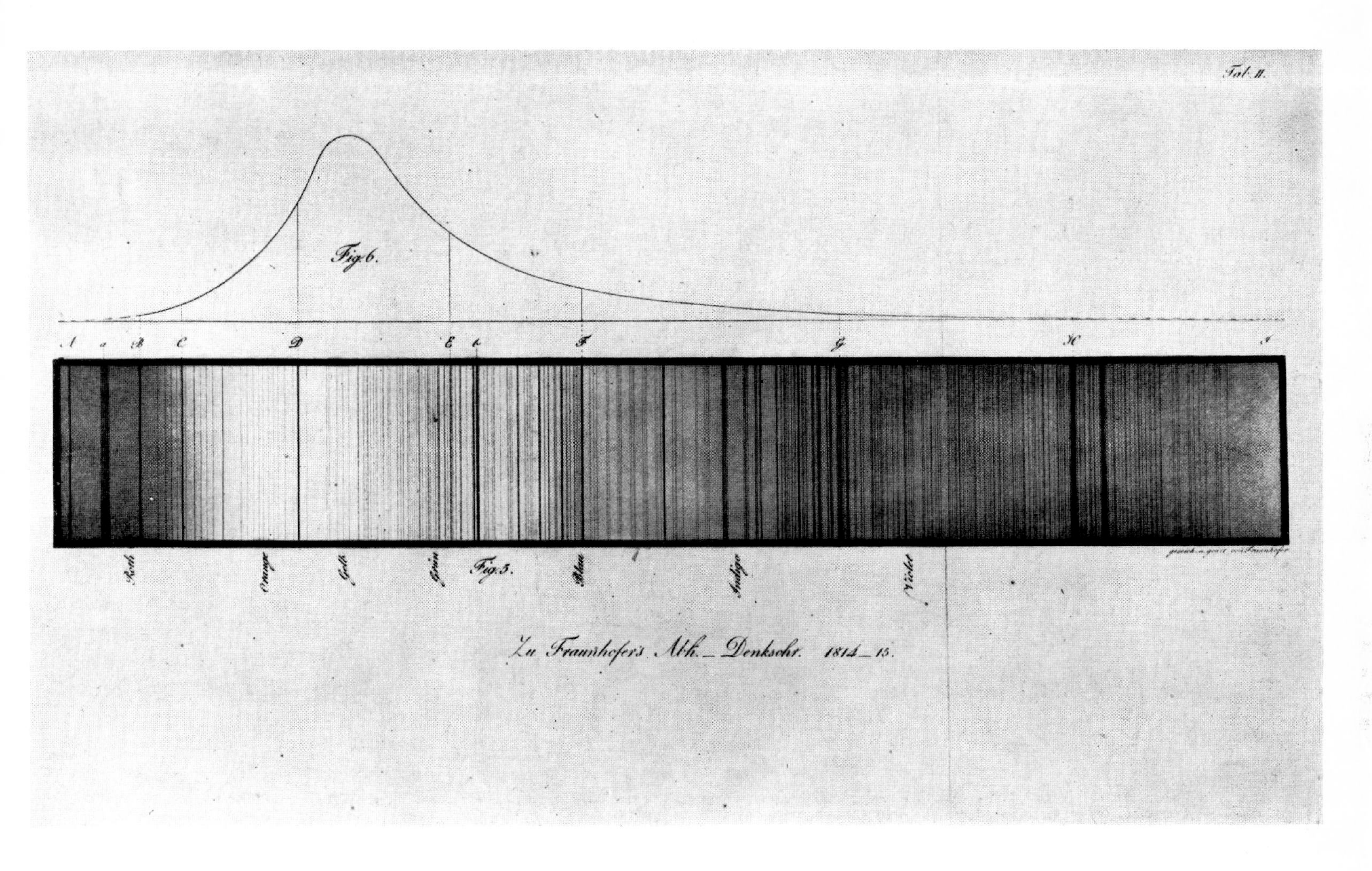

43 This picture of Fraunhofer lines is taken from an original work by Fraunhofer himself. These lines appear as narrow dark stripes on the variously colored background of a continuous spectrum. The upper curve indicates the brightnesses which the human eye can identify in the solar spectrum. The yellow components thus appear as the brightest.

nesses of the members of this system already indicated an extension in depth of the system of fixed stars. Herschel concerned himself with the identification of this phenomenon.

Herschel directed his excellent reflecting telescope in turn at each of several thousand points in this sphere and he counted the stars which could be identified in each case in the field of view, which covered an area about a quarter of the size of the full moon. Assuming certain conditions, the number found had to indicate the extent of the stellar system in the particular direction, a larger number of stars implying a greater extension in depth. From these "star gauges," Herschel constructed the picture of a lenticular area around the Sun filled with millions of stars. We now know that such a picture does not correctly reproduce the actual conditions in our stellar system, the Milky Way. The results of the counts and thus of dimensions derived from them were subject to the influence of interstellar dust, which, to a varying extent depending on the different directions and distances, screens off the light coming from the stars. Even if this is disregarded, Herschel could not have reached the dim stars, marking the most remote areas of the system, with the telescope at his disposal. Nevertheless, Herschel's method of research has retained its validity.

The investigation of individual objects such as the stars, star systems, or matter more diffusely spread in Space on the one hand and comparative observations of entire areas of sky on the other became fields of study for practical astronomers. Both of these tasks called for the use of powerful telescopes, in the first case so that more and more details could be seen, or to identify dimmer objects, and in the second to reach out farther into Space. This is why

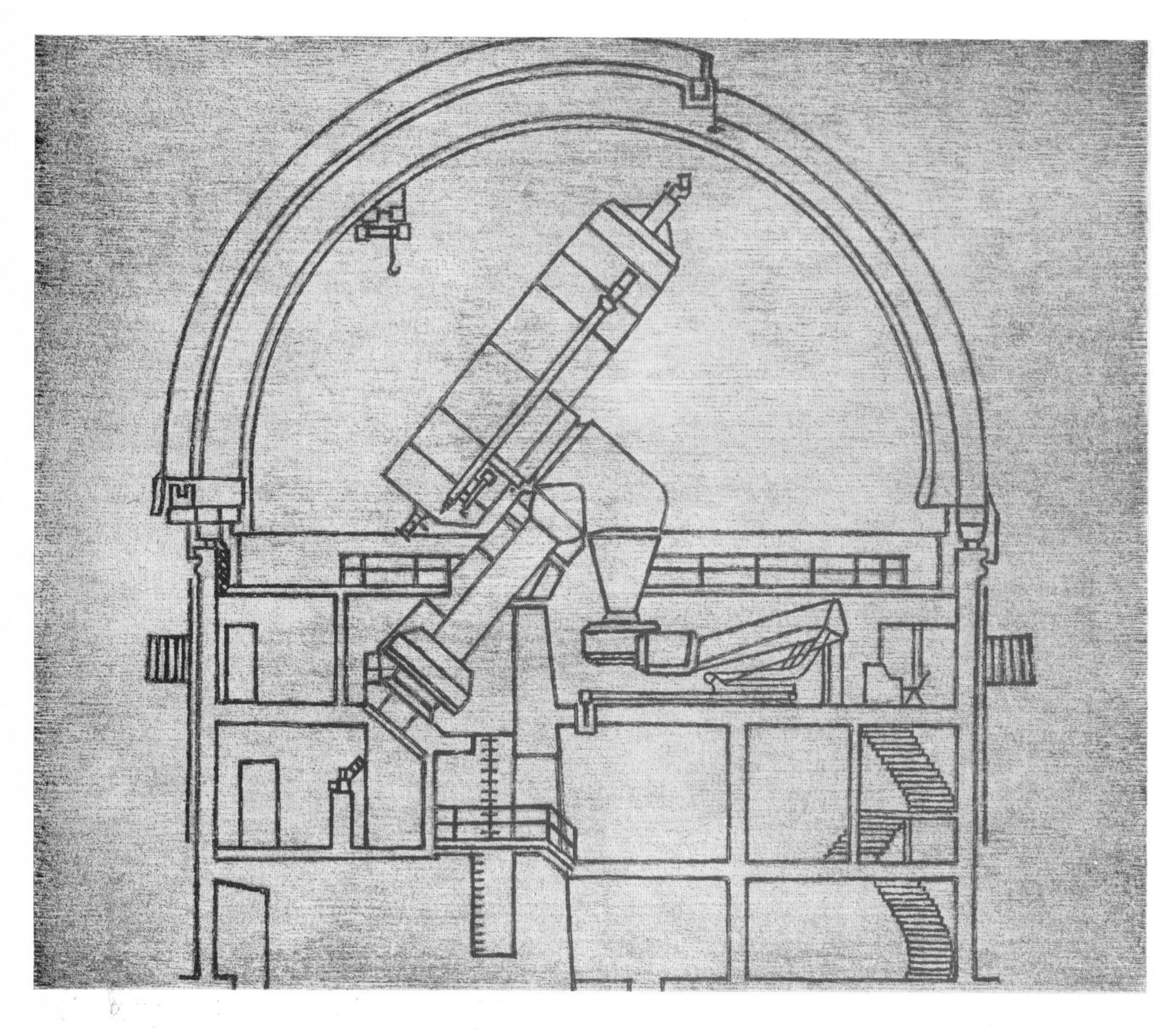

44 The 2-meter (79-inch) reflector of the Ondrejov Observatory of the Czechoslovak Academy of Sciences rests on concrete pillars. Two vertically adjustable platforms permit convenient operation of the instrument which is housed in a revolving dome of 20 meters (65 feet) in diameter. The Coudé spectrograph is accommodated in a room under the floor of the dome and can be seen on the left of the drawing. Irrespective of the position of the telescope, the light from the celestial body to which the telescope points is directed into this room via a total of four mirrors.

45 This is the dome of the Ondrejov Observatory of the Czechoslovak Academy of Sciences. It houses a 2-meter (79-inch) reflector.

the great reflecting telescopes are a characteristic feature of our century. Only instruments of this kind have the large aperture required. While investigations were concentrated on individual objects, the question of the field of view was of no importance. On the other hand, with the Schmidt mirror system and other new optical arrangements, reflectors with large fields of view were now available.

The great reflecting telescopes

The first big telescope of recent times, the 60-inch reflector on Mount Wilson in California, was specifically designed for spectroscopic use. Even before it was completed in 1908, work had begun on its larger "brother," the 100-inch reflector, likewise on Mount Wilson. Both telescopes have made a great contribution to the history of astronomy and even today are still in full use. The plans for the latter instrument overlapped in turn those for the Hale telescope on Mount Palomar in California, which was named for George Ellery Hale, the initiator of the project. The inauguration ceremony of this technical masterpiece with its 200-inch mirror took place in 1948, the final commissioning of the instrument in 1950.

The 20-ton disk of Pyrex, the material selected for the mirror, was successfully cast at the second attempt in 1936. After a further eleven years, including the interruption caused by the Second World War, the task of grinding the mirror was completed. These two decades were a period of intensive planning, design, and construction which confronted all those involved with problems of an entirely new kind.

46 The Astrophysical Observatory of Potsdam was one of the first observatories in the world to be established with an expressly astrophysical orientation. This spectrograph was part of the excellent equipment of the institute. The upper hemispherical part encloses the prisms that break up the light into its variously colored components, deflect the rays and direct them on to a photographic plate which can be positioned on the lower left of the instrument in the picture.

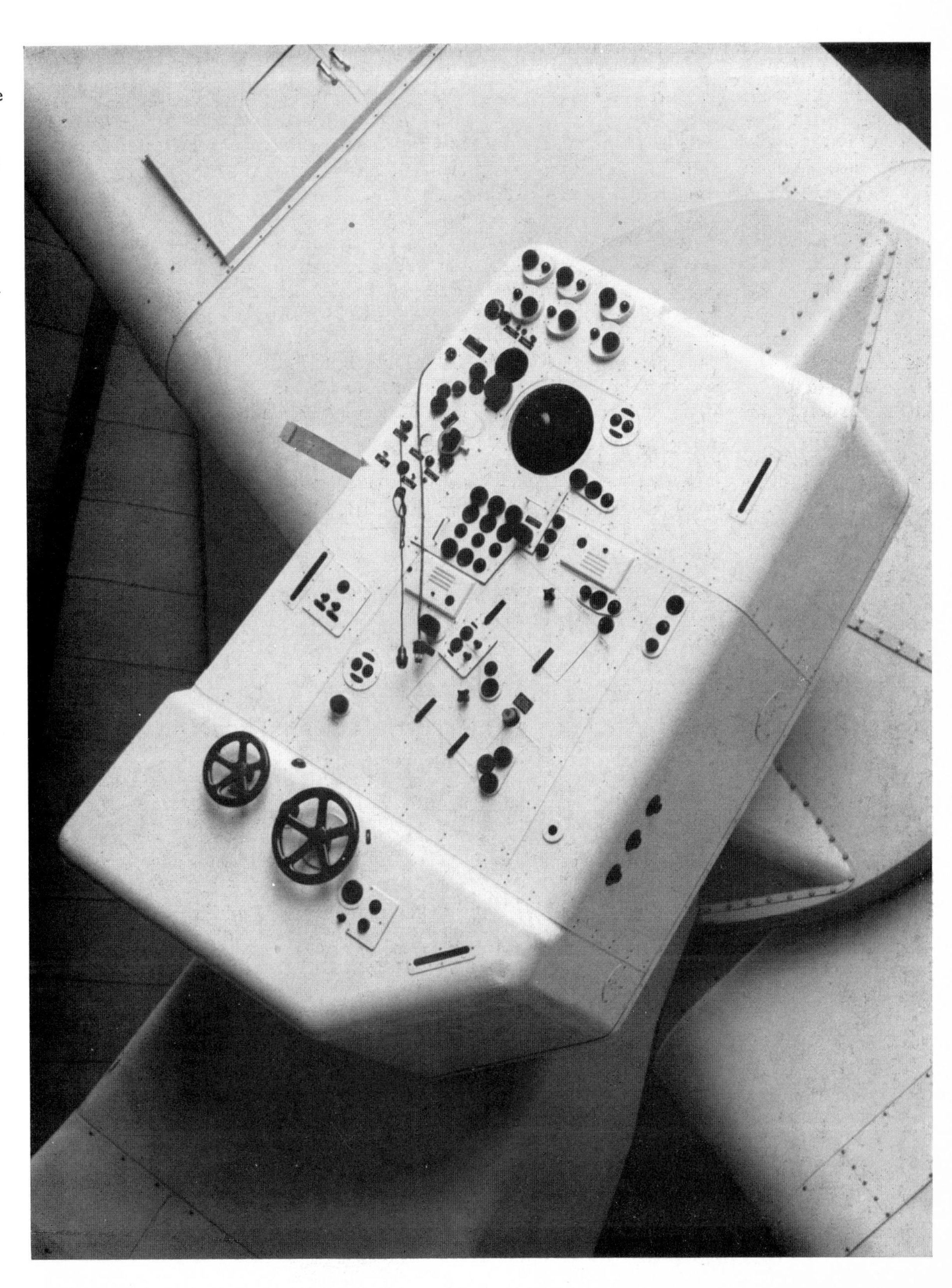

47 Weighing 1 1/2 tons, the spectrograph on the 2-meter (79-inch) universal reflector at the Karl Schwarzschild Observatory, Tautenburg, is an impressive example of the technical advance achieved in auxiliary equipment. A large number of controls are necessary to ensure that the spectrograms obtained meet the required standards of quality. Thermostats maintain the interior of the spectrograph at a constant temperature.

It may be remarked that in the construction of large astronomical instruments the dimensions of the individual components in general are of the same order as those usual in heavy engineering, but the precision required is of an incredibly exacting degree and becomes even more critical with every increase in dimension. It is obvious that this precision must be maintained whatever the position of the telescope tube in relation to the horizon and despite the great fluctuations which may occur in the temperatures at which the instrument is operated. Courage and optimism in the way in which they tackled these problems characterized the group of workers led by Hale. Originally—at the end of the 1920's!—they had planned a 300-inch telescope but had then limited themselves, mainly for financial reasons, to a mirror diameter of 200 inches.
At the time that the Hale telescope was inaugurated, there were eleven instruments with a mirror diameter of more than 60 inches in the world, but by 1964 this number had doubled. It might be imagined that this would have been enough to meet the needs of astronomers, but this was not so. In 1964, the Academy of Sciences of the USA published a ten-year program which called for more giant telescopes, including the three instruments of the 140-inch (3.5 meters) and 200-inch (5 meters) classes. The report stressed the need as well for a study to ascertain how large an instrument could be built and also stated in particular that the installations in the Southern hemisphere, which has been relatively neglected with respect to optical observations, should be increased. This would permit observation of objects which can only be seen in southern latitudes, by far the most important of these being the two Magellanic Clouds which are the nearest star systems. Furthermore, full use could be made of the excellent climatic conditions found in certain southern locations. Since then the construction of giant telescopes has been started in various parts of the world and

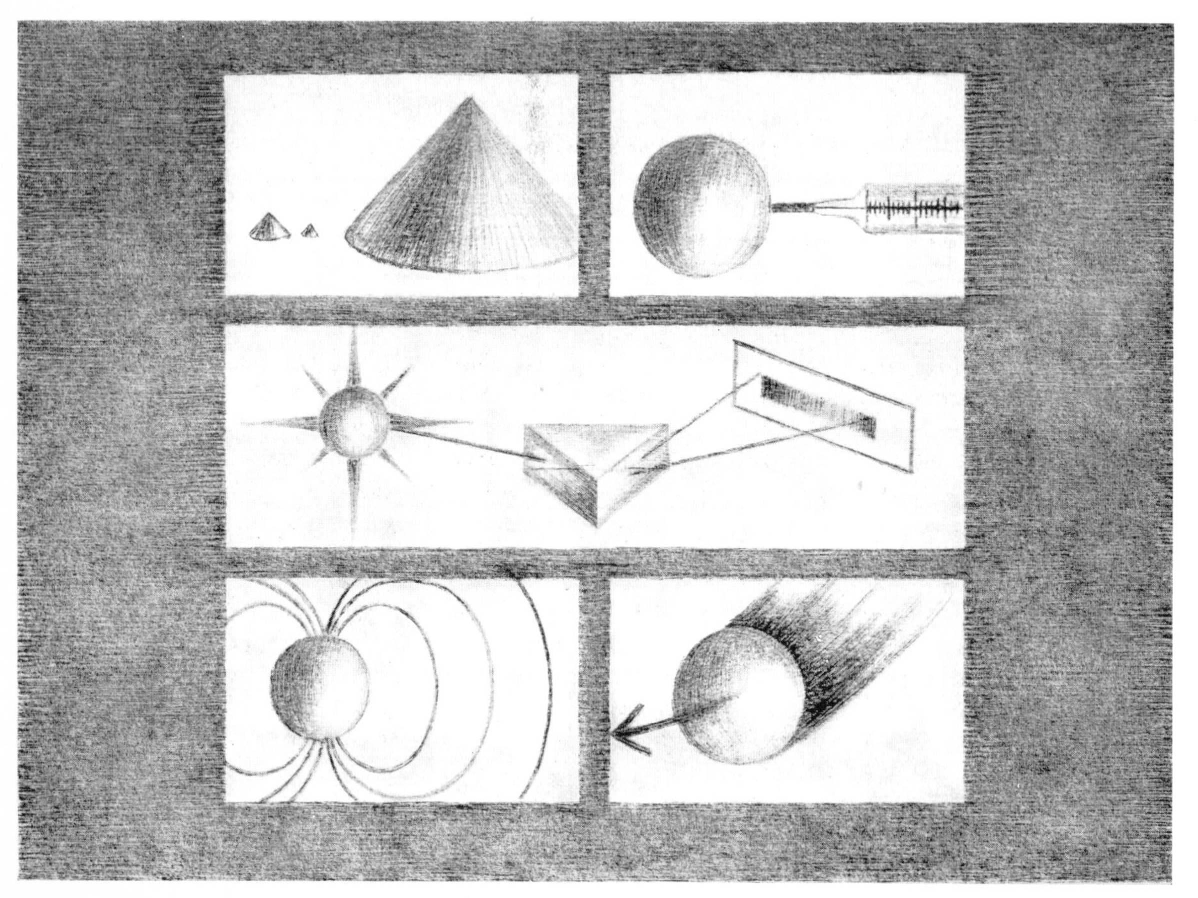

48 Splitting up the light coming from celestial bodies to obtain a spectrum (center) enables conclusions to be drawn about the physical states (upper right) and the chemical composition (upper left) of the layers emitting radiation. It is also possible to determine the presence and strength of magnetic fields (lower left). Above all, however, the components of motion lying in the line of sight can be measured as radial velocity (lower right). In certain cases, it is even possible to ascertain the rotational speed of the object in this manner.

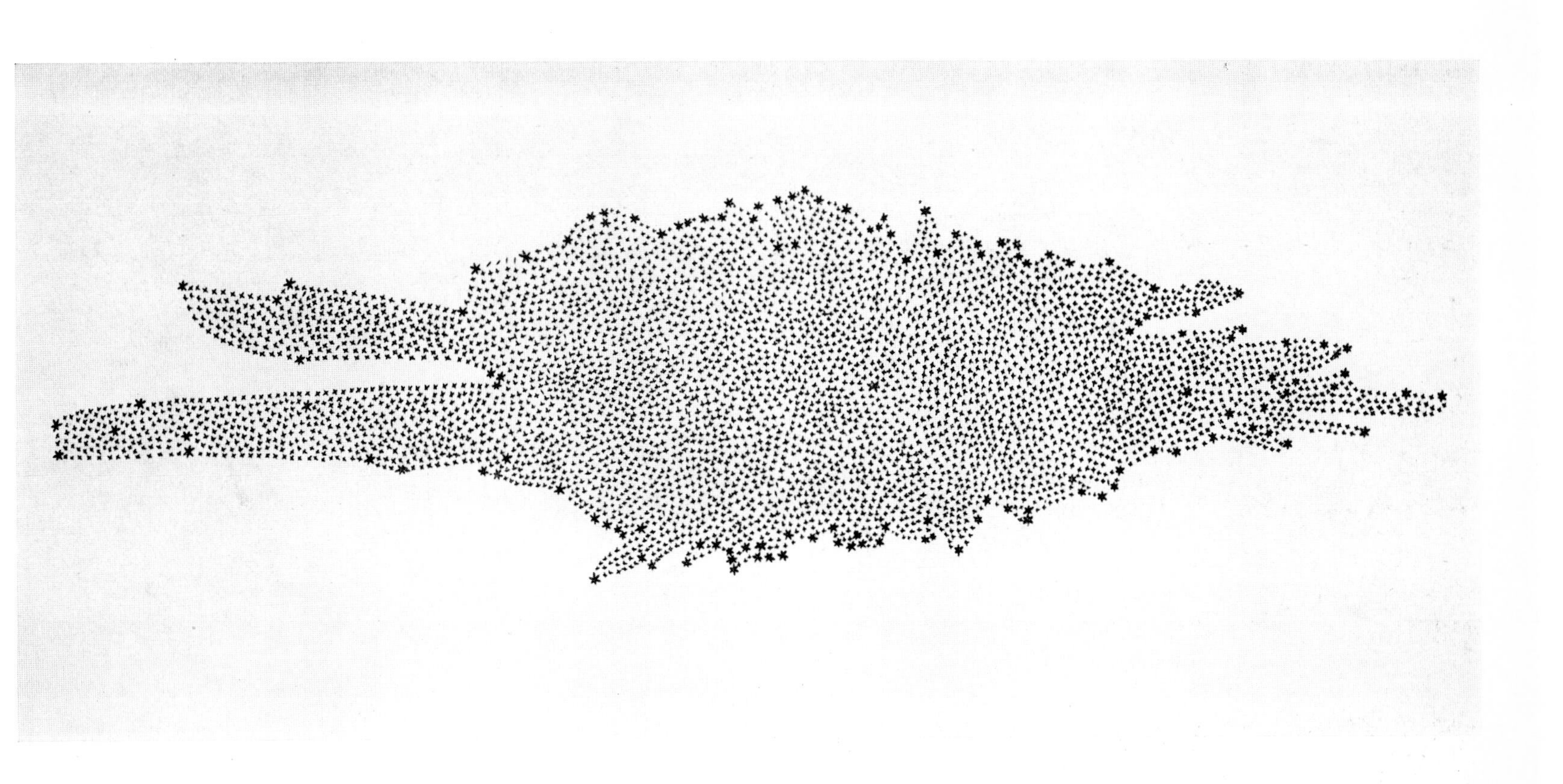

49 William Herschel presented the result of his star-counts in diagrammatic form and wrote "... that the stupendous sidereal system we inhabit, this extensive stratum and its secondary branch, consisting of many millions of stars, is, in all probability, *a detached Nebula*." [8]

projects already under way have been completed. The largest optical telescope in existence (at present undergoing trial operation) is the 6-meter (236-inch) reflector of the Academy of Sciences of the USSR. The unconventional mounting of this instrument is apparent at first sight. An innovation which will certainly influence future constructions is the fact that the basic plane of movement is aligned according to the horizon and not, as is usual in optical astronomy, according to the celestial equator. It is true that the movements of the telescope are complicated by this, but the difficulties associated with the great masses moved are considerably reduced by the layout chosen, which is symmetrical in relation to the force of gravity. The telescope was built in Leningrad, the grinding of the mirror being carried out in a workshop specially built for the purpose in Moscow. To convey an impression of the mighty scale of the instrument, it may be noted that a 70-centimeter reflector is mounted on it as an auxiliary instrument for sighting and checking the correct functioning of the main instrument. On one of the observation platforms of the fork there is a spectrograph reflector of 2 meters diameter.

Installations such as this certainly come within the sphere of large-scale research. It is therefore clear that cooperation extending over national frontiers is now more important than ever before in astronomy, too. Thus, for some years the European Southern Observatory (ESO) on La Silla (2,400 meters altitude) in Chile has been operated as a joint undertaking by six countries of Western Europe. A 3.6-meter telescope is being built there. Another project of this kind is the Anglo-Australian plan for a 3.75-meter (150-inch) reflector to be erected in an observatory in southeast Australia.

Photography and radiation reception

Despite the tremendous improvements in telescopes, the achievements of modern astronomy have only been possible through simultaneous developments of means for receiving radiation. The small items of accessory equipment, which record or process the incoming light, are scarcely less important than the mighty telescopes which pick up, collect and pass the radiation to the receiving device. Above all, in the present age, when the construction of observation instruments for optical astronomy is approaching

226 *Dr.* HERSCHEL's *Catalogue of a ſecond Thouſand*

This method of viewing the heavens ſeems to throw them into a new kind of light. They now are ſeen to reſemble a luxuriant garden, which contains the greateſt variety of productions, in different flouriſhing beds; and one advantage we may at leaſt reap from it is, that we can, as it were, extend the range of our experience to an immenſe duration. For, to continue the ſimile I have borrowed from the vegetable kingdom, is it not almoſt the ſame thing, whether we live ſucceſſively to witneſs the germination, blooming, foliage, fecundity, fading, withering, and corruption of a plant, or whether a vaſt number of ſpecimens, ſelected from every ſtage through which the plant paſſes in the courſe of its exiſtence, be brought at once to our view?

WILLIAM HERSCHEL.

Slough near Windſor, May 1, 1789.

Firſt Claſs. Bright nebulæ.

I.	1785		Stars.		M.	S.		D.	M.	Ob	Deſcription.
94	April	28	61 Urſæ	f	0	6	n	2	17	2	cB. pL. E ſp nf. vgmbM. 3½ l. 2′b.
95	—	—	—	f	35	0	n	2	7	2	cB. cL. E. np ſf. bM. 4′l. 3 b.
96	May	1	14 Canum	f	5	30	n	1	12	2	vB. cL. mE. ſp nf. ſmbM. 6′l. 1′½ b.
97	—	—	—	f	7	58	n	0	47	1	vB. pL. E. nearly mer. gmbM.
98	—	—	—	f	36	50	ſ	0	12	1	cB. pL. R. vgmbM.
99		—	27 (γ) Bootis	p	13	46	ſ	1	46	2	vB. S. R. vſmbM.
100	Sept.	10	41 Ceti	f	13	43	n	0	48	1	cB. pS. R. mbM. See III. 431.
101		—	67 —	p	17	19	n	0	25	2	cB. pL. E. near mer. mbM. 5′ l.
102	—	—	—	f	21	37	ſ	0	13	2	cB. pL. R. mbM.
103		24	14 Delphini	p	16	10	ſ	0	3	1	vB. L. gmbM. er. beautif. object.
104		28	93 (Ψ) Aqua	f	1	8	n	0	42	1	cB. cL. E. near. mer. gmbM. F. rays.
105	Oct.	3	47 Ceti	f	26	24	ſ	0	37	1	cB. pL. iR. mbM.
106		—	89 (π) —	f	38	10	ſ	1	24	2	cB. cL. iR. bM. 3′ dia.
107		6	20 Eridani	f	4	3	ſ	1	4	2	vB. R. BNM. 1′½ dia.
108		8	111 (ξ) Piſcm	p	34	22	ſ	0	1	1	cB. vL. iR. p. vBſt.
109		26	12 Eridani	p	7	17	n	2	54	3	cB. pS. lE. mer. mbM. r. 1′½ l.

110

50 This excerpt from the catalog of new nebulae and star clusters by William Herschel explains why the researcher, despite the short span of human life, can venture to make statements about cosmic development. Herschel compares the Universe to a garden in which the plants may be seen simultaneously at different stages of their growth and decline, thus presenting, at one glance, an impression of the entire sequence of their existence. This is the "spatial concurrence" of a "chronological succession" which still remains the key to cosmic development.

51 A look in the great dome on Mount Palomar reveals the massive Hale telescope and its interesting design.

So that the 500-ton instrument can be safely handled, the tube is carried in a fork which can turn in two forced-lubrication bearings. The great "horseshoe" supporting it at the northern end also allows the telescope to be directed towards the area of the celestial pole.

52 The observer in the lift is on his way to work. He will enter the cabin at the front end of the Hale telescope and will "drive" in the instrument to the position in the sky which he is to investigate. Despite its large diameter, the cabin only obscures about ten percent of the area of the mirror.

53 Work can begin. The observer sits in the cabin in the Hale telescope and inserts the slide in preparation for a photograph. For the whole of the exposure period, he will check the automatic movement of the telescope through the eyepiece. There is provision for using accessory instruments at other points along the optical path, e.g., in a room for work at the coudé focus. In this case, the instrument is operated directly from the points in question.

54 Installation of the 6-meter (236-inch) reflector of the Academy of Sciences of the USSR in its gigantic dome. It can be turned together with the horizontal circular base plate. The second movement required for covering the whole of the visible hemisphere of the sky is around an axis that passes through the upper wheel.

55 The observatory of the Academy of Sciences of the USSR with the 6-meter reflector is situated near Zelenchukskaya, a small village in the northern Caucasus about 80 kilometers (50 miles) to the northwest of Mt. Elbrus at a height of 1,800 meters (5,900 feet) above sea level.

56 The European Southern Observatory in Chile is equipped, among other instruments, with a 1.52-meter (60-inch) reflecting telescope for spectrographic observations. The three fans which can be seen on the tube ensure a controlled circulation of air to reduce the turbulence in the tube, giving a very much better image of the stars.

technical and financial limits, it is essential that the optimum use is made of incoming radiation.
This is why the first photograph of a celestial body marks the beginning of a new era in astronomy. In 1839, Daguerre, the inventor of the photographic process named for him, took a picture of the Moon. The final breakthrough in the application of photography to astronomy then followed about two decades later when more sophisticated techniques and materials became available. Apart from this, improvements in the mechanical systems driving the telescopes became necessary, and it was only when telescopes followed the slow rotation of the starry sky more accurately that good exposures, taken over a period of several minutes or more, became possible. Until this time, when only visual observations had been made, no particular attention had had to be paid to this point. After further improvements of this kind, the vast field which now confronted astronomers was very rapidly appreciated.

In a report on photographic work by the Harvard College Observatory in the USA dating from the middle of the last century, it is stated that "... each night, in fact, opens new vistas ..."[5] and "There is nothing, then, so extravagant in predicting a future application of photography to stellar astronomy on a most magnificent scale."[6] This expectation was confirmed but on a scale undreamt of. When exposed, the photographic plate captures all the objects seen in the field of view of the telescope. Its ability to pick up the action of light in a cumulative manner thus enables it to surpass the human eye in range. Expressed in rough terms, this means that a 30-minute exposure will show stars of only 1/100th the brightness of those which can be seen by the human eye through the same telescope. In the case of extended objects, structures can be identified which would otherwise escape the observer. Objects of different degrees of brightness appear as different grades of photographic density. Because brighter stars produce

57 The area of our immediate cosmic surroundings is fairly uniformly filled with stars. When, for his observations, the astronomer takes an area in the sky of about the size of the full moon, the stars he sees occupy a conical area, opening out into Space (central part of the illustration). The size of the optical system—eye, binoculars or telescope—collecting the light determines to a major extent the range of observation, i.e., the depth of the cone. However, this also fixes the number of stars which are accessible to the observer (lower part of the illustration). When seen with the naked eye, this number is small but increases greatly when optical aids are used.

spots of larger diameters on the photographic emulsion, it is also possible to draw conclusions about their brightness, permitting photometric evaluations to be made. In cases of doubt, the exposed plate can be subjected to further examination. Comparisons with photographs taken at a later date also provide objective evidence of changes in the sky. From this there developed a wide-ranging field of activity for photography, from the recording of changes in the brightness of stars and parallax measurements to the checking of conclusions drawn from Einstein's theory of relativity. Furthermore, photographic emulsions are sensitive to radiation on certain wavelengths which can only be seen very poorly or not at all by the human eye.

As a whole, photography has had an incalculably fruitful effect on astronomy and has probably a greater share in its advancement than in any other science. And yet, especially since the Second World War, it has been necessary to share the observation time available for photometric work with another process, namely, photoelectric brightness measurement. With this, the light falling on a photoelectric receiver arranged at the focal point of the telescope generates an electric current which provides an indication of the incoming amount of light. This process is a quantitative one in the real sense of the word, whereas the eye and the photographic plate react only in a qualitative manner. If one of two stars is twice as bright as the other, the photoelectric cell will produce exactly double the current. The eye and the photographic plate, on the other hand, initially show only a difference in brightness, the actual result only being obtained after calibration or comparative observations.

An additional advantage is that the light-sensitive cathode, the receiving part of the cell, is able to make better use of the incoming light than the photographic emulsion. With the latter, only one percent of the quantity of light falling on it acts on the film under the most advantageous conditions whereas a figure of up to 20 percent is attained with modern photoelectric cathodes. This means that the photoelectric method gives still greater range and, above all, greater accuracy. It is evident that in the Space Age and the "remote control" of astronomical measuring instruments that this implies these photocells are particularly convenient aids since it is relatively easy to code the currents carrying the data and to transmit them to the Earth.

In the case of terrestrial photometric work, photography has so far been able to maintain its position since a single exposure can record a multiplicity of objects and provide data for further processing. With photoelectric observations, on the other hand, it was necessary to measure all the stars included in the program one after the other. However, photography has recently met with serious competition in the form of video amplifiers and image storage tubes, the latter being extensively used for television techniques. These combine the image-producing qualities of the photographic plate with the advantages of the photoelectric cathode. These new developments will replace photographic techniques in some branches of astronomical observations. There are still certain drawbacks to be overcome, but work is in progress on these problems and the prospects of success are good.

Astronomical data processing

The great advances achieved in the sector of actual observation have also had an effect on astronomical data processing, the process which directly follows the work of observation. The wealth of information being obtained can only be handled in the short time available, if it is to retain current validity, by using automated processes for the phase of data reduction. The necessity for this is obvious when it is realized that a photographic plate from the Schmidt Telescope of the Karl Schwarzschild Observatory in Tautenburg, when used for a 30-minute exposure of a star-filled part of the sky, will capture the images of several hundred thousand stars. If this exposure is to be used for brightness measurements, there is an enormous difference between telescope time and reduction time, even when only small samples are taken. Efforts must be made to keep the latter component as low as possible by utilizing high efficiency "measuring machines." Without such aids, the accumulation of further plate material would very rapidly become senseless and ultimately would nullify the work carried out with the telescope.

The technical solutions to this problem are varied and have attained varying stages of perfection. In the most advanced designs, the work of the measuring unit is controlled by a computer. Similar problems are arising in spectroscopy. Here, the quantity of details visible in a spectrum of medium to high resolution likewise calls for automatic processes in which the densities of the photographic plates can be

58 Sophisticated mechanical and optical systems and electronics are relieving astronomers of the tedious routine work in the evaluation of celestial photographs. "Galaxy," an automatic measuring unit, supplies in one hour the accurate positions on the plate and the data equivalent to stellar brightness for 900 objects. This unit was developed in collaboration with staff members of the Royal Observatory, Edinburgh.

converted into measured variables and certain unwelcome factors eliminated. The characteristics of the numerous absorption lines, for instance, could then be given in the form of numerical values that would provide quick information and be suitable for further processing.

Questions of automatic operation also play a great role in the work carried out directly with the telescope during the night. Here, it is above all a question of the optimal utilization of the valuable telescope time. Experiments on a medium-size telescope have demonstrated that the supervision and control of the sequence of work by an electronic computer coupled to the instrument can double the number of stars observed per night. This increase in efficiency was obtained with an investment representing only a small percentage of the cost of the complete telescope.

Apart from the function mentioned, the computer can also be used for an immediate initial processing of the data obtained by observation. It thus also contributes to a better utilization of observation time. Consequently it is obvious that all the big telescopes are now being equipped at the planning stage or later with equipment of this kind.

A role which is at least of equal importance is played by the computer in the subsequent stage of the theoretical interpretation of the results and in the purely theoretical research work. Here, it relieves people of the task of making comprehensive and purely mechanical calculations and, in fact, it is only a "service" such as this which enables many theoretical problems to be solved at all. Included here are calculations relating to the internal structure of stars, their development in time and the reconstruction in model form of the emergence of a star system. The saving in time possible can be demonstrated by an historical example. The method of calculation for determining the orbit of the planet Mars from the observations available, exactly as specified by Kepler, was recently programmed for a modern computer. The machine then took only eight seconds to work out the same problem for which Kepler needed four years!

Exploration of new spectral areas

It was the beginning of a new epoch in astronomy when human beings succeeded in overcoming certain limitations of their sensory organs and of their terrestrial location. The exploration of new spectral regions is the special characteristic of this phase.

Space is filled with electromagnetic waves covering a broad range of wavelengths. The human eye reacts only to radiation of very specific wavelengths—to *light*, a radiation process with a wavelength of a few ten-thousandths of a millimeter. With certain detec-

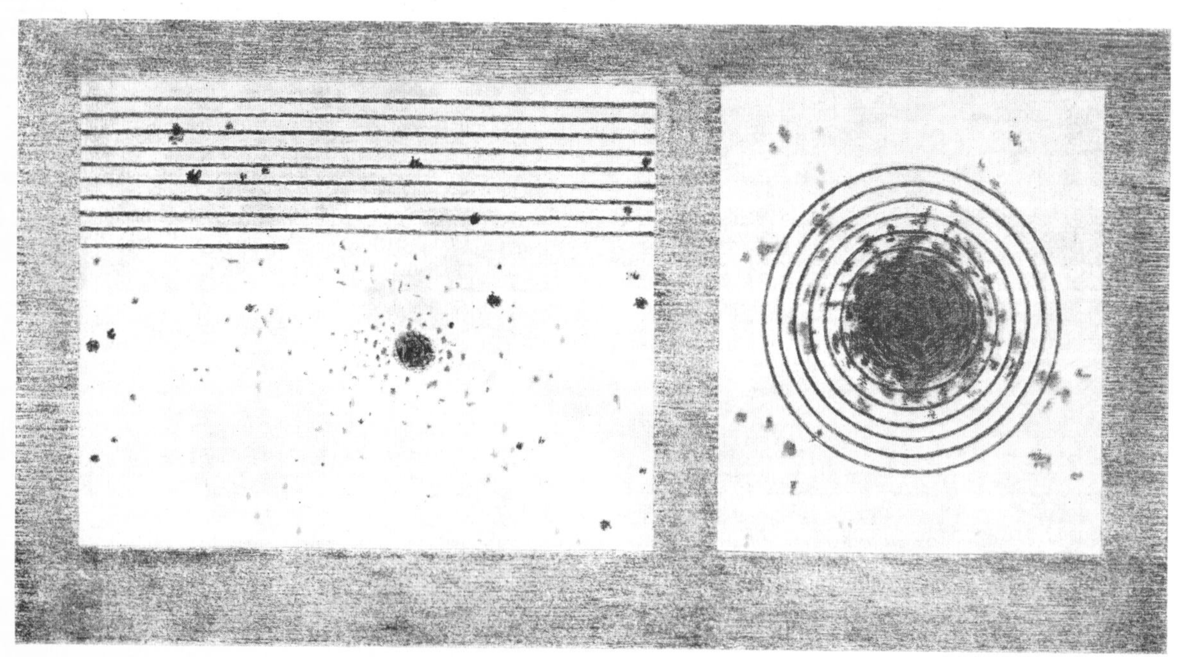

59 With the "Galaxy" automatic unit, the process of measurement is divided into two steps. First of all, the selected area of the photographic plate (left) is scanned line by line and the positions of the stars noted in the course of this are retained in the storage register of an electronic computer. In the actual process of measurement which then follows, the individual images of the stars are scanned once more, one after the other, in a circular pattern (right). From the information thus obtained, the computer establishes corrected positions to an accuracy of one-thousandth of a millimeter and determines the brightness of the objects.

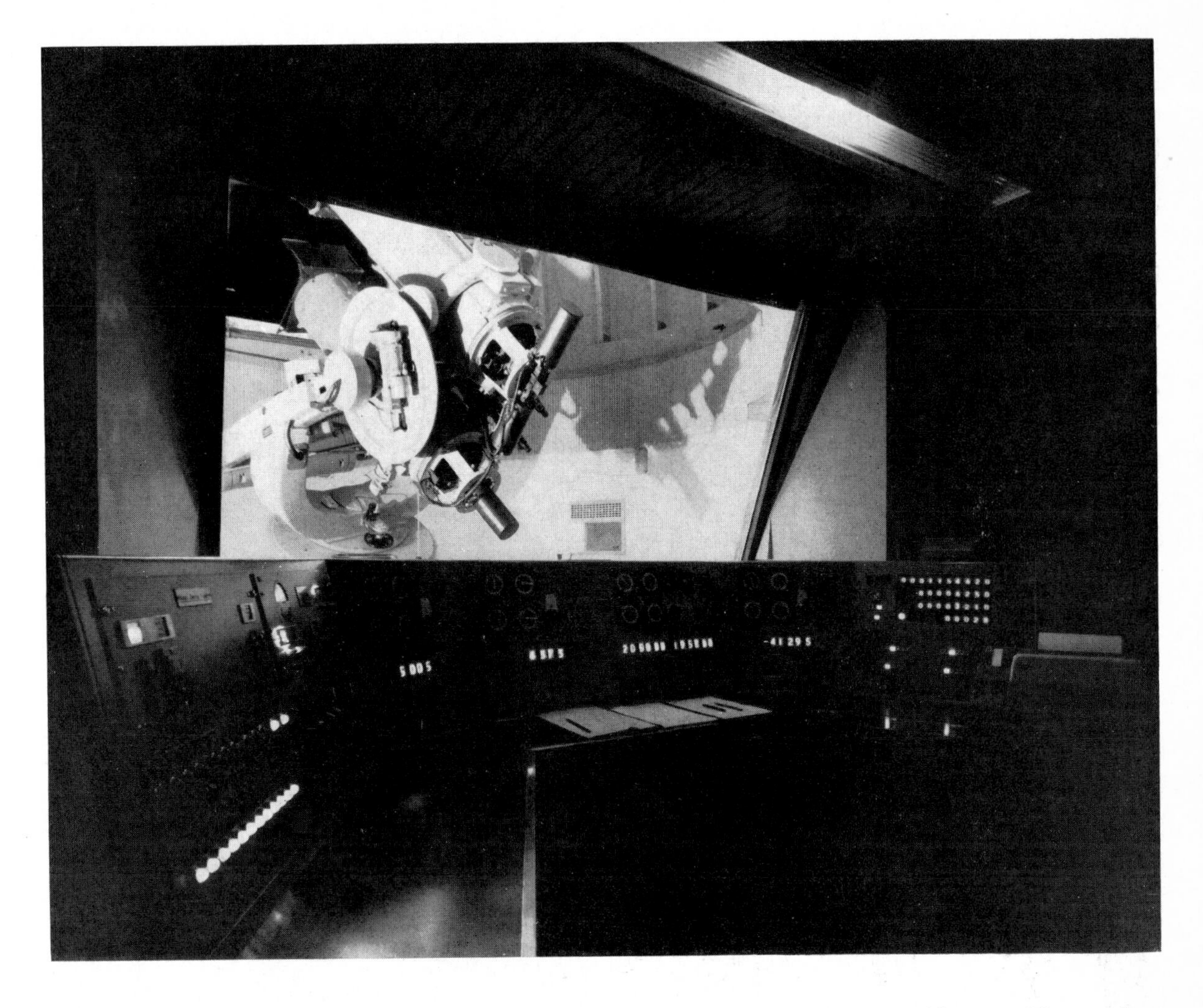

60 These two 40-centimeter (16-inch) reflecting telescopes of the Royal Observatory, Edinburgh, work automatically. They are monitored and remote controlled from a room separate from the dome.

tors and to some extent with photographic plates as well, it is possible to identify infrared radiation, which is "similar" to light, and ultraviolet radiation. The investigation of the centimeter to meter wave-range of cosmic radio wave radiation opened up a completely new field. At the beginning of the 1930's, while working on radio interference problems, the American engineer K. G. Jansky discovered a source of interference which obviously lay outside the atmosphere and was ultimately located in a certain part of the Milky Way. However, it was only after the Second World War that real progress began to be made in radio astronomy. Nowadays, one can speak of it as an independent branch of astrophysical research. The physical processes in Space produce radiation on all wavelengths. It is clear that our understanding of these processes has become very much greater with the increase in the range of wavelengths accessible to us. This applies, for example, to our knowledge of the causes of radiation in the Cosmos. New findings point to previously unsuspected sources drastically different from the "... gently burning stars of classical astrophysics."[7]

Like the visible spectrum, the radio spectrum also contains discrete lines. Uninterrupted by interstellar dust, these can be traced throughout the entire star system and have supplied data which would have remained inaccessible if only visual means of observation had been available. With the discovery of the radio emission of interstellar molecules, including those of organic substances, radio astronomy has taken us a step further in the investigation of the origin of life as well.

The radio astronomer uses a number of items of equipment to carry out his investigations. Particularly familiar are the great dish-shaped reflectors which, like an optical reflecting telescope, gather the radiation and pass it to a receiver. These installations are inferior to optical instruments in the sense that their power of resolution is much less, and fine details of structure become "blurred" in the observations. This can be overcome by linking up the information picked up by these receiving aerials in a special way. This restores the superiority of the radio astronomer since he can now make out details in the radio sources of the order of a thousandth of a second of an arc. To take the example used before, this would correspond to the diameter of the coin when seen at a distance of 4,000 kilometers or 2,500 miles. But it must be noted that achievements of this kind necessitate the use of the most advanced equipment. Atomic clocks are essential so that the information obtained with two radio telescopes can be compared with an accuracy of at least one-millionth of a second and combined as a single measurement. What is particularly necessary, however, is the worldwide cooperation of astronomers so that we can really speak of "intercontinental radio astronomy." An example of this sort of cooperation is a "bridge" which was set up in 1971 between the largest radio telescopes of the USA and the Soviet Union.

Finally, it may also be mentioned that radar techniques are being applied to astronomy. Initially, it was only the case that experience gained in the design of radar technical equipment influenced the design of radio telescopes.

Today, however, the radio waves transmitted and received by radio telescopes after reflection from the bodies of the solar system and man-made space vehicles are providing a great deal of information about our immediate cosmic surroundings. Radar has thus become one of the few but extremely important methods with whose assistance the astronomer can establish direct contact with the objects he is researching.

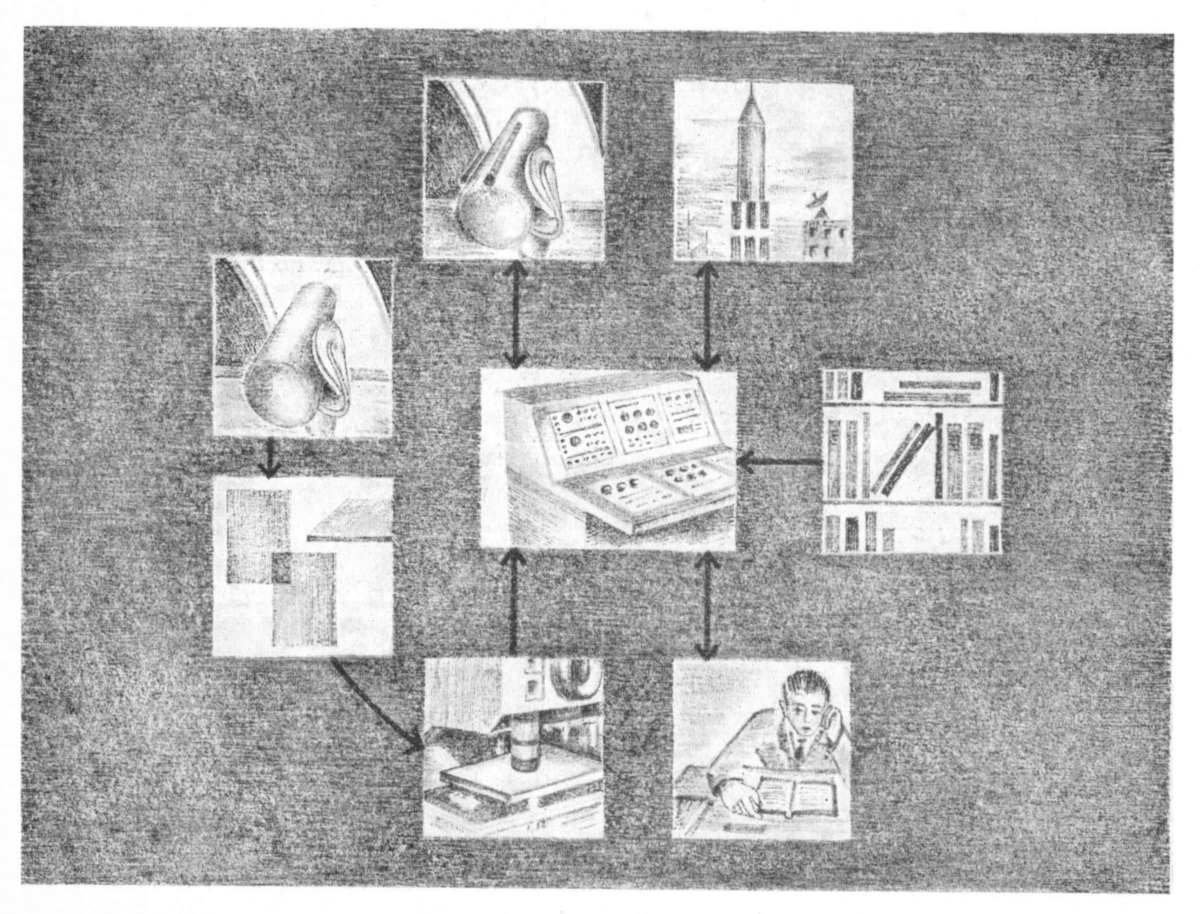

61 Computers are indispensable for the whole of present-day astronomical work. They control telescopes (upper left) and evaluate the material obtained, possible in the form of photographic plates (left and lower left). Computers are essential for the successful launching of satellites (upper right), and theoreticians are obliged to use them for their extensive calculations (lower right). The contents of entire catalogs and card index systems can be stored in them, freeing the user not only from tedious searches but also from statistical evaluation work (right).

62 In 1951, a theoretically predicted line of the radio spectrum of interstellar hydrogen on a wavelength of 21 centimeters was discovered independently at three different spots in the world.
In Holland, this was done with a 7.5-meter (288-inch) radio telescope of the "Würzburg-Riese" type.

63 The largest fully maneuverable radio telescope in the world is this instrument which was put into service in 1971 by the Max Planck Institute for Radio Astronomy in Bonn. Screened from terrestrial radio interference to a large extent, this giant with a reflector of 100 meters (3,937 inches) in diameter stands in a small valley in the Eifel Mountains. Constructed of steel and about the size of a football field, it can be moved freely to pick up radio radiation down to a wavelength of about 1 centimeter.

64 With the radio telescope, one cannot speak of "seeing" in the strict sense of the word. If, however, the intensity of the radiation picked up by the instrument from the individual points in the sky is plotted in a curve, an idea is obtained of how the "radio sky" appears on the wavelength in question. This chart made by an English observatory covers an apparent area of one hundred times that of the Moon and shows the positions of numerous small sources of radio radiation. The broad, curved lines are due to interference effects.

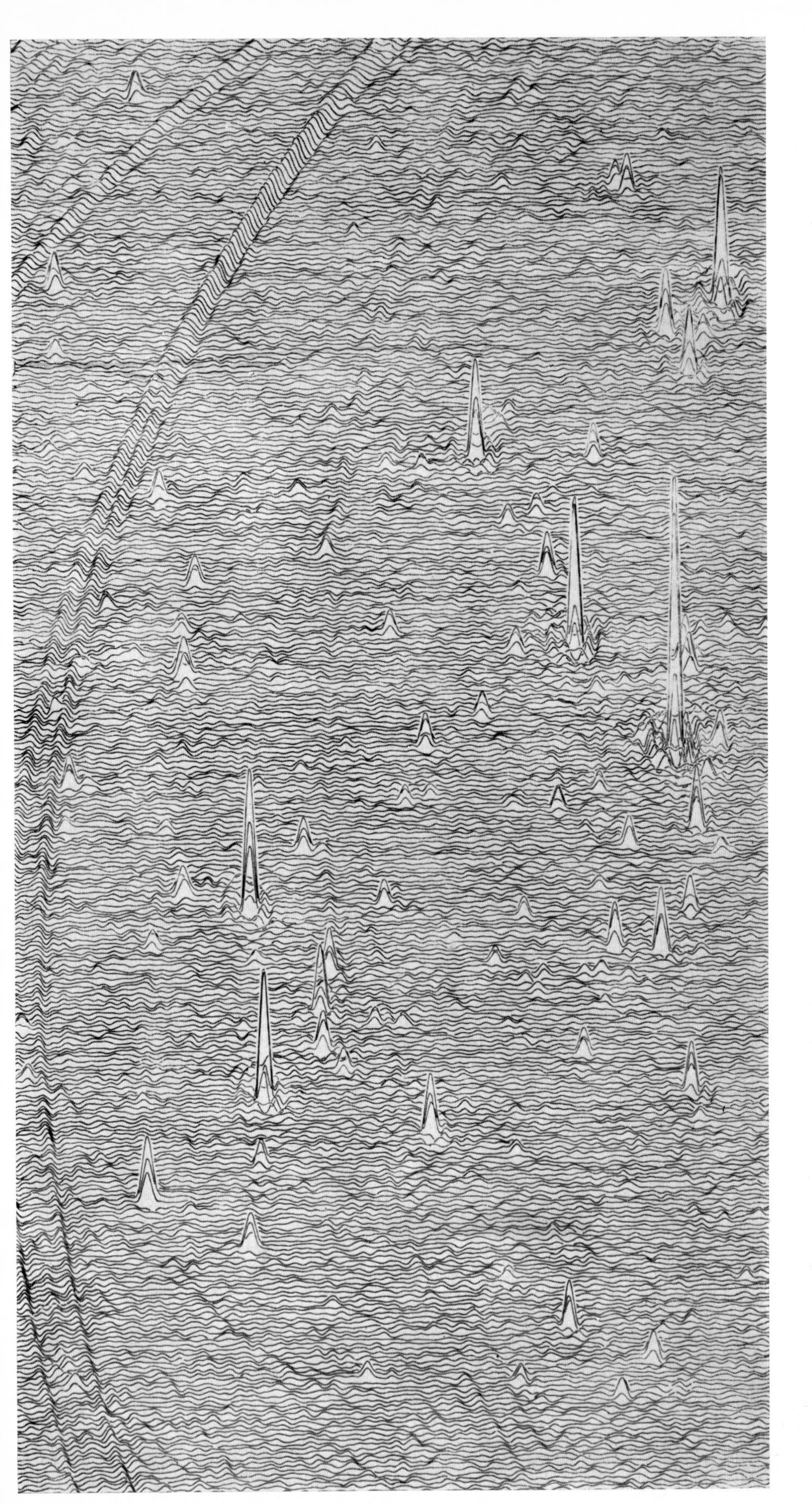

Radio waves are found in the longwave area of the electromagnetic spectrum, but investigations of the radiation components with wavelengths shorter than those of light have also produced important new findings. Shorter wavelengths are associated with ultraviolet, X-ray and gamma radiation emitted by celestial bodies and interstellar gas. These wavelengths cannot be measured by terrestrial instruments, however, since the atmosphere of the Earth absorbs shortwaves. It is consequently necessary to send the measuring instruments to considerable heights above the Earth by using satellites, for instance. Even observers in the optical spectral ranges have a basic interest in "observation platforms" of this kind, since the terrestrial atmosphere frequently impedes astronomical observations. The resolving power theoretically possible is restricted by the to-and-fro movement of the stars known as directional scintillation. The brightness in the night sky, poor visibility and, in all probability, the increasing cloudiness of the sky, which are the consequences of modern civilization, cannot be excluded either.

Space probing satellites

Extraterrestrial observation points can be established or approximated in various ways. High-flying aircraft have been successfully used for collecting data, and an important role is still played by observations from balloons, chiefly on account of the long observation times which these permit. But the ceilings of balloons are limited, and there remain experiments which are inconvenienced even by the residual atmosphere still present at those heights. The only other alternatives are the high altitude rocket, which has the disadvantage of only brief measurement times, and the ideal but costly artificial satellite as vehicles for instruments.

Satellites with astronomical missions have been launched by both of the great space nations. Astronomical satellites have been included in the Soviet Kosmos series, and the Interkosmos satellites, a joint project of the socialist countries, have also carried out tasks of this kind. The specific purpose of the Soviet satellite Prognoz was the investigation of the gamma and X-ray radiation from the Sun and also its corpuscular radiation, known as the "solar wind." The satellite had a markedly elliptic orbit which took it half-way to the Moon and thus permitted measurements in areas at some distance from the Earth. The most comprehensive astronomical measurement program so far was carried out by the American satellite OAO 2, an orbiting astronomical observatory. The objective of this space vehicle, which was launched as long ago as the end of 1968, was the observation of the stars in the ultraviolet spectral area not accessible from the Earth. A scrutiny of the entire sky also formed part of the undertaking, four telescopes equipped with television cameras supplying information about some 5,000 stars. Another part of the program carried out exact photometric measurements on numerous individual objects. The principal instruments used for this were six 20-centimeter reflecting telescopes. In this experiment, full use was made of the satellite's capability for programmed operations. Its great flexibility permitted the selection of objects which promised data of exceptional importance. More than 400 instructions per day were transmitted to the control system of the vehicle, which then investigated the objects in the precise order specified by the terrestrial control station.

Since the summer of 1972, astronomers have had at their disposal for measurements in the ultraviolet spectral range a large instrument, an 82-centimeter reflecting telescope, carried by another OAO satellite. Launched on approximately the fifth centenary of the birth of the great Polish astronomer, Nicholas Copernicus, this satellite bears his name.

Projects of this kind have provided a new insight into the state of the "hot" stars. There was previously considerable vagueness about the structure and development of these objects, since only a small part of the energy emitted by these celestial bodies can be observed from the Earth. On account of the high temperatures of the radiating strata, the greater part of this energy falls within a spectral region which is almost entirely absorbed by the terrestrial atmosphere.

In view of the success and potential of extraterrestrial observations, it must be asked whether there is any point at all in the further development of Earth-bound

65 Two radio telescopes, separated by continents, act as a single gigantic instrument when a coordinated evaluation of the data they pick up is carried out. In the future, it will be possible to achieve a further rise in efficiency by linking up a terrestrial radio telescope with another instrument stationed on the Moon.

66 Due to the influence of the envelope of air around the Earth, only visible light and the radiation in the radio wave range can reach the Earth. Infrared radiation, whose wavelength is between these two, is halted a few kilometers away from the Earth, except where "wavelength windows" allow it to pass through. Very shortwave radiation from Space is kept even farther away. Very long radio waves are reflected away at the ionosphere. For the observation of the radio components that do not reach the surface of the Earth special aids such as balloons, rockets and satellites are used.

67 This view from the observatory on Mount Wilson over the West Coast of California with Los Angeles, Hollywood and other urban areas gives an idea of how much astronomical work is disturbed by the increase in terrestrial light sources.

68 A Soviet spaceship of the manned "Soyuz" type on the launching pad in Baikonur, north of Lake Aral, ready to go into orbit around the Earth.

69 Last checks of the astronomical orbital observatory OAO 2 before its launching at the end of 1968. The solar cells on the "wings" supply the spacecraft with the electricity it needs.

70 The satellite "Copernicus" was launched as the fourth and last space vehicle in the OAO series. This orbiting observatory carries out astronomical observations with the aid of an 82-centimeter (33-inch) reflecting telescope, whose great accuracy of alignment permits the spectrographic investigation of individual stars. The research program began with ultraviolet measurements on a very young star and continued on objects which are known as sources of X-rays but have not yet been in every case visually identified.

astronomy. The answer to this is quite definitely in the affirmative. It cannot be a question of competition between these two fields; it is rather the case that they have to be used to supplement each other as far as possible. In this way, an ambitious investigation of the Cosmos is being carried out from different points of view. The tasks which can be solved by terrestrial observations are so diverse and numerous that they surely cannot be handled entirely by the far more costly satellites or space laboratories. It should be considered that the investments called for in the ten-year program for optical and radio astronomy (to which reference has already been made) amount to one percent at most of the present expenditure by the USA on space research. For the cost of a single Type OAO satellite with a service life of one year, it is possible at the present time to build several 5-meter (200-inch) reflecting telescopes. The useful life of these instruments, however, is a few decades. The potential represented by observatories equipped with instruments such as these cannot be ignored even in the Space Age.

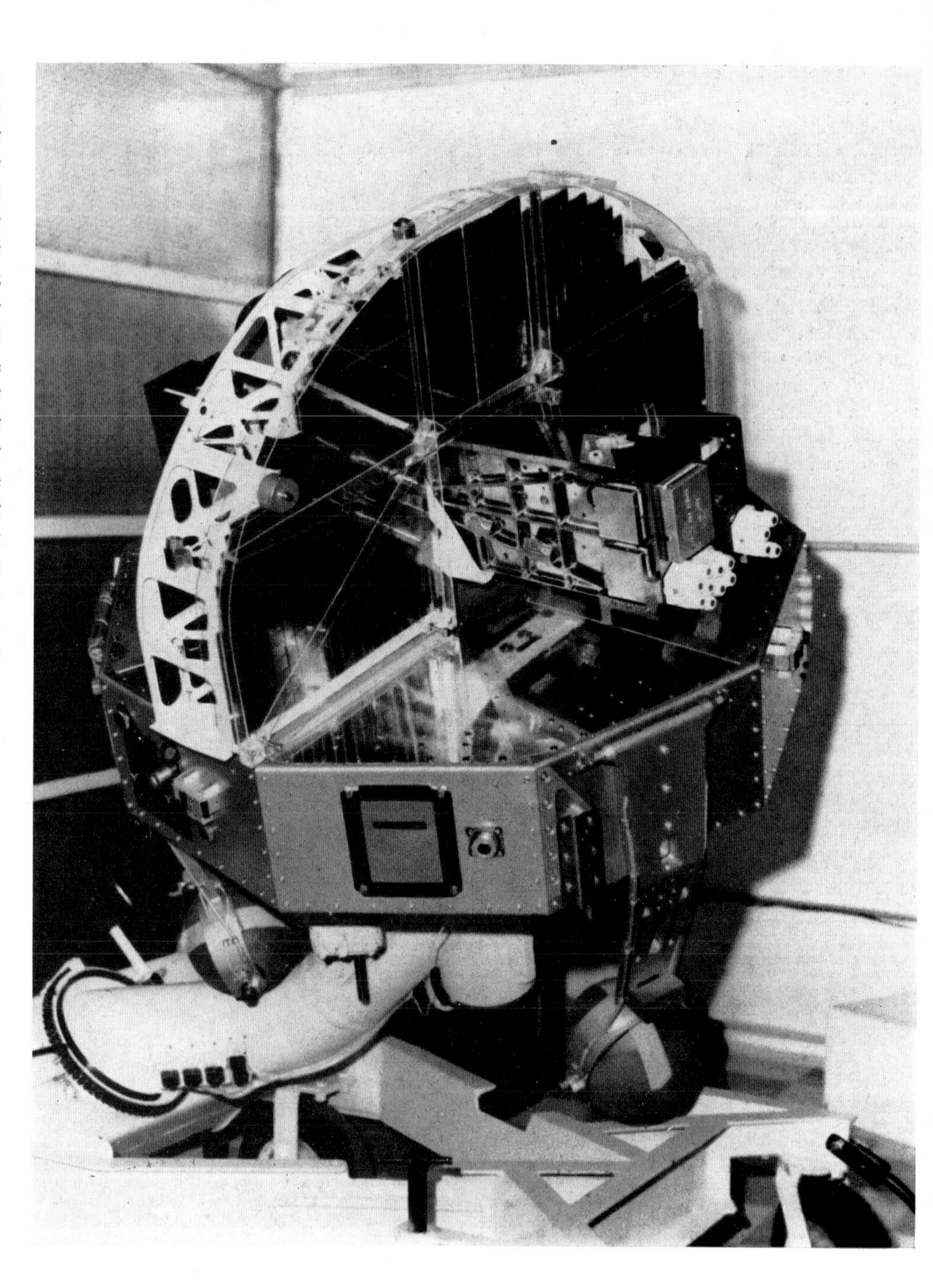

71 The Sun, together with its various influences on the Earth, is also the object of extraterrestrial investigations. A whole series of satellites—the OSO series—was launched from Cape Kennedy in a wide-ranging research program to "get closer" to the dominant star of our part of the universe. These satellites consist of two parts. The disk underneath rotates very rapidly and acts as a gyroscope to stabilize the position of the vehicle in Space. The rotation is controlled

by gas jets installed in the balls on the outriggers which are still in the lowered position. The upper part, the "sail," is always kept facing the Sun so that it can carry out measurements.

72 Successful rocket experiments opened up a new field of research —"X-ray astronomy"—in 1960. Today, work is being carried out in this branch by, among other things, a series of special satellites which supply observation data.

The picture shows Explorer 42, the first satellite made by the USA and launched by another nation. It was put into orbit by an Italian team from a floating platform off the East Coast of Africa.

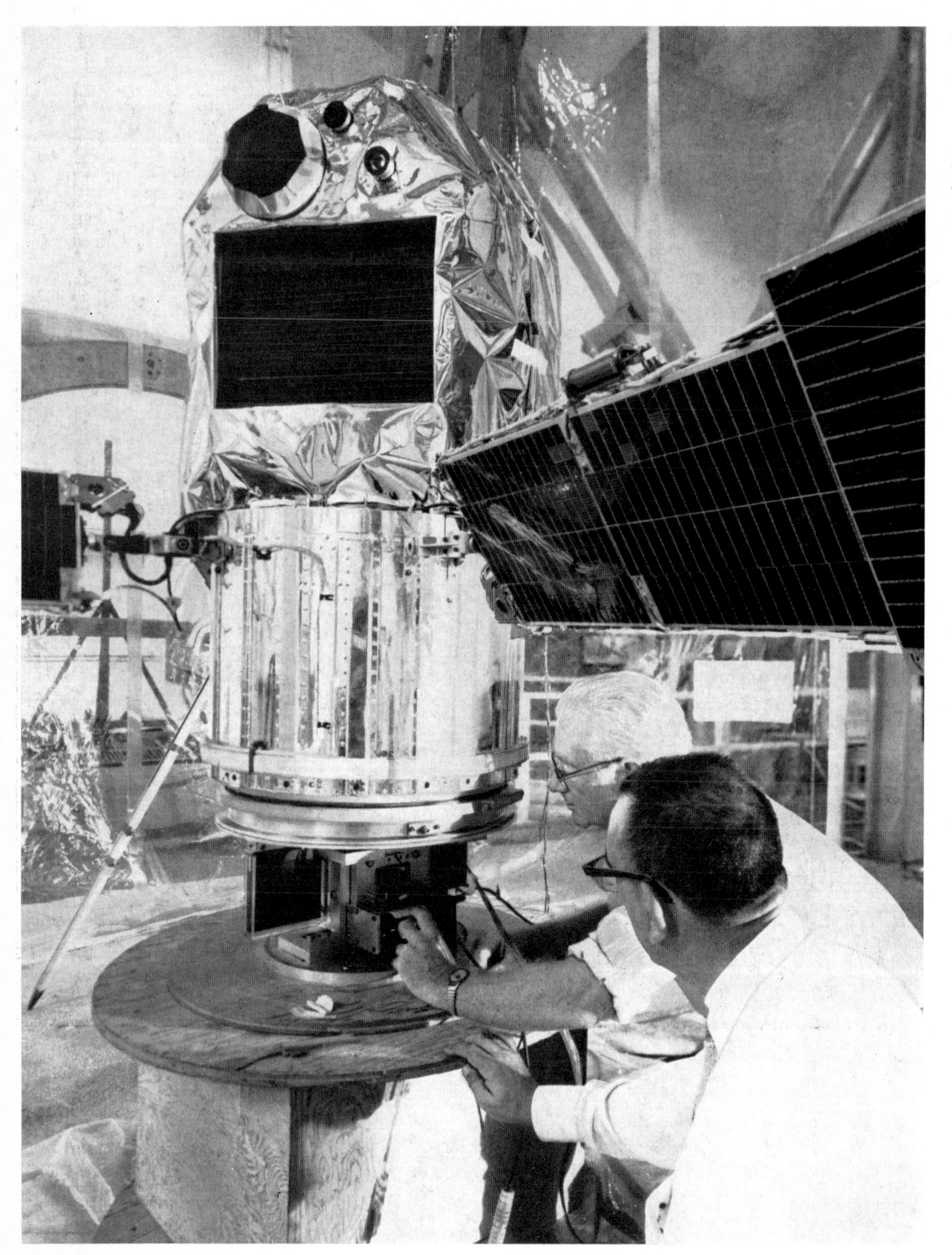

73 Explorer 42 in the laboratory. Through the round apertures visible at the top, solar and stellar sensors control the position of the satellite in Space. The large dark rectangle below them is part of the actual measurement section, the X-ray telescope. The four big paddles carry the solar cells, the power plant of the satellite.

74 The present state of space technology permits the use of large vehicles with a service life of several years as extraterrestrial laboratories. These help to solve research problems in the most diverse branches of science including astronomy. The launching of the Soviet spacecraft "Salut" in April 1971 was a major step forward toward orbital stations of this kind. Initially unmanned, it was approached by the Soyuz 10 and 11 spaceships which then, in turn, linked up with it. The three-man crew of Soyuz 11 stayed up for 22 days in the Salut-Soyuz station. Its overall length in the coupled state was 20 meters (65.6 feet), its maximum diameter was four meters (13 feet) and it weighed about 27 tons.

75 Apart from the coupling of a Soviet space vehicle to an American one for a joint mission planned for 1975, the Skylab project is regarded as the most important undertaking in U.S. space activities now that the Apollo Moon landing program has been completed. Skylab is a space laboratory that provided accommodation and working facilities for several crews in succession. The astronauts stayed up for up to 84 days and worked on more than 50 experiments. These included medical and biological research and extensive technical investigations. Astronomy profits above all from the observation of the Sun on different wavelengths without interference from the terrestrial atmosphere. The picture shows how the surface of the Earth can be scanned from the space vehicle, permitting the visual location of mineral resources.

Chapter III
Gas in Space

Of major importance for the development of modern astronomy was the change, initiated at the beginning of the Nineteenth Century, in the direction of research which had been followed for many centuries. Even though celestial mechanics continued to dominate astronomical activities at this time and in the following period, individual astronomers and physicists started to concern themselves at this stage with questions of the material composition and physical state of the celestial bodies. Observations revealed that the surfaces of the Sun, the stars and most of the luminous nebulae consisted of gas. In the meantime it has been positively established that the cores of stars are likewise of gas. Since most of the matter in Space is concentrated in the stars, this means that at the present time the greater part of cosmic matter is in a gaseous state.

An increasing amount of astronomical work is being carried out in the field of cosmogony, a subject which embraces all the questions referring to the formation of the celestial bodies, for instance the planets, the stars and the clusters, which consist of thousands of millions of individual stars and are known as galaxies. Even the evolution of the Universe, cosmology, is a subject which has been and is being researched with some success.

The Twentieth Century will probably go down in the ancient history of astronomy as that period in which the first significant advances were made in the clarification of cosmogonic and cosmological problems.

It is part of the intrinsic nature of the investigations that they are very largely theoretical, i.e., their findings are based on a logical application of the universally valid laws of Nature. If these theoretical investigations are to have a wider meaning, it is essential that their findings be compared with the actual state of affairs. This is a constant source of new tasks for astronomical observation. These consist not only in the collection of observational data of even greater accuracy, but also in the confirmation of new effects which have been theoretically predicted. Only through this close interplay between theoretical and practical astronomy is it possible to acquire a more profound understanding of the processes of development in Space.

Energy sources of the stars

In most stars, which from a terrestrial point of view consist of hot spheres of gas of gigantic size, a kind of self-acting control mechanism has enabled pressure and temperature relations of such an order to become established that the disintegration or expansion of the stars is prevented for fairly long periods of time. They constantly emit a great stream of energy, which is maintained by a "power station" in the core of each star.

Once the magnitude of the energy emitted by the Sun had been fully appreciated, astronomers wanted to know more about the energy sources of the stars. Of the hypotheses formulated with reference to this, the one advanced by Julius Robert Mayer (1814—1879) subsequently became the best known. In his view, the energy lost by the Sun through radiation can be replaced by a constant influx of mechanical energy. The essential condition for this is that the Sun is struck by an unceasing stream of fairly small solid bodies from the solar system. He considered that this produced energy of a largely kinetic kind, which was then converted into heat and partially radiated by the Sun. The critical confrontation with the hypothesis advanced by Mayer has revealed, however, that it is incompatible with basic facts of observation.

The work of Hermann Helmholtz (1821—1894), the physicist who was also active in the field of physiology, is regarded as the first scientific attempt to explain the nature of the energy sources of the Sun. He based his assessments of this energy on gravity, the subject which was then most familiar to physicists since it was precisely in the field of mechanics that the greatest theoretical advances had been achieved. The point of departure for Helmholtz's considerations was the knowledge that the contraction of an expanded ball of gas causes mechanical—in this case potential—energy to be converted into other forms of energy, e.g., thermal energy. In this way, a ball of gas with the mass of the Sun could develop heat through continuous contraction and, with a part of the thermal energy generated, supply energy by radiation for a period of about 30 million years. Although this span of time appeared adequate to astronomers and physicists at first, others, including geologists, doubted whether it was long enough for the development of life on the Earth.

Progress in clarifying the question of stellar energy sources first became possible with the discovery of radioactivity by Henri Becquerel and Marie and Pierre Curie at the end of the last century. Knowledge of the decay times of certain radioactive isotopes of chemical elements enabled the age of the Earth's

crust to be determined fairly accurately. Indirectly, it was also possible to gain an idea of the age of the Sun, which recent findings estimate at about five thousand million years. This is about one hundred times more than the figure for the age of the Sun arrived at by Helmholtz under the conditions already described. From this it is evident—and this conclusion may be applied to all the stars—that the energy released by the contraction of a cloud of gas of large size and mass cannot be the sole source of stellar energy if a star such as the Sun is able to maintain its radiation of energy at a roughly constant level for a period of several thousand million years.
The further efforts of astronomers and physicists to unravel the mysteries of the energy sources of the stars were closely associated with progress in nuclear physics. It became increasingly apparent that these energy sources could only be of an atomic nature.
In the 1920's, theoretical evidence was obtained that under the conditions of extremely high temperature and density prevailing in the core of stars sufficient atomic nuclei are able to overcome the natural electrical barriers surrounding them and to achieve fusion. This results in the formation of fairly heavy chemical elements with a fraction of the mass of the atomic nuclei involved in the reaction being converted into another form of matter, radiation. The realization of this was at least the foundation for the understanding of the principal source of energy of the stars.

The Sun

At the end of the 1930's, the two physicists Hans Bethe and Carl Friedrich von Weizsäcker described a complete cycle of nuclear reactions in the course of which four hydrogen nuclei are converted into one helium nucleus with a simultaneous release of energy. In the following period not only was a constant improvement achieved in the atomic data needed for nuclear reaction, but it also proved possible to identify other important nuclear reactions taking place in the release of energy in the stars.
For a number of reasons, the Sun occupies a special position among the stars. The distance between it and the Earth is so short in astronomical terms that it is the only star in Space on which surface details can be identified. The Sun also represents the center of gravity of the solar system. From the gravitational effects exerted by it on the planets, for instance, it is possible to work out its mass with the aid of Newton's law of universal gravitation. If it were possible to place the Sun in one pan of a pair of celestial scales, the other pan would need to hold a mass 333,000 times that of the Earth in order to balance it. Just as impressive is the magnitude of the stream of energy constantly emitted by the Sun, for which the term luminosity is used. The luminosity of the Sun is so great that it exercises a major influence on the energy balance and thus on the physical state of all the bodies in the solar system. Each square centimeter of the surface of the Sun emits a quantity of energy equivalent to 6.4 kilowatts. To underline the implications of this figure, it may also be said that to maintain the flow of energy passing into Space from the interior of the Sun about 4,300 million kilograms of matter must be converted into radiation every second. In view of this impressive "slimming course" of the Sun, the question naturally arises of how long this process can be kept up without affecting its internal constitution. Even though this appears to us to be a tremendous loss of weight, in comparison with the total mass of the Sun it is still so tiny that it will only become noticeable after about ten thousand million years.
For reasons associated with the physical processes in the movement of radiation energy—which also includes the light we are able to register with our eyes—through atoms, it is impossible to "see" the interior of the Sun with conventional optical or radio methods of astronomical observation. Only a layer about 500 kilometers (300 miles) thick and known as the photosphere is accessible to direct observation. When it is realized that the Sun has a diameter of about 1,390,000 kilometers or 864,000 miles, this means that we can scarcely look under its skin. What can be seen in actual fact? The first thing that is apparent on examination of photographs taken through telescopes of the entire disk of the Sun is that it is not so bright at the edge as at the center. This darkening at the edge is explained by the fact that one can see further into the Sun in the center of the disk than at the edge. Since the inner parts of the Sun are hotter, they therefore emit more radiation. Thus there is an increase in the observed brightness at the center of the disk where the line of sight penetrates more deeply.
Close examination of the photosphere reveals what is termed the granulation of its structure. The appearance of the bright and dark elements in this structure changes from one minute to the next. Spectro-

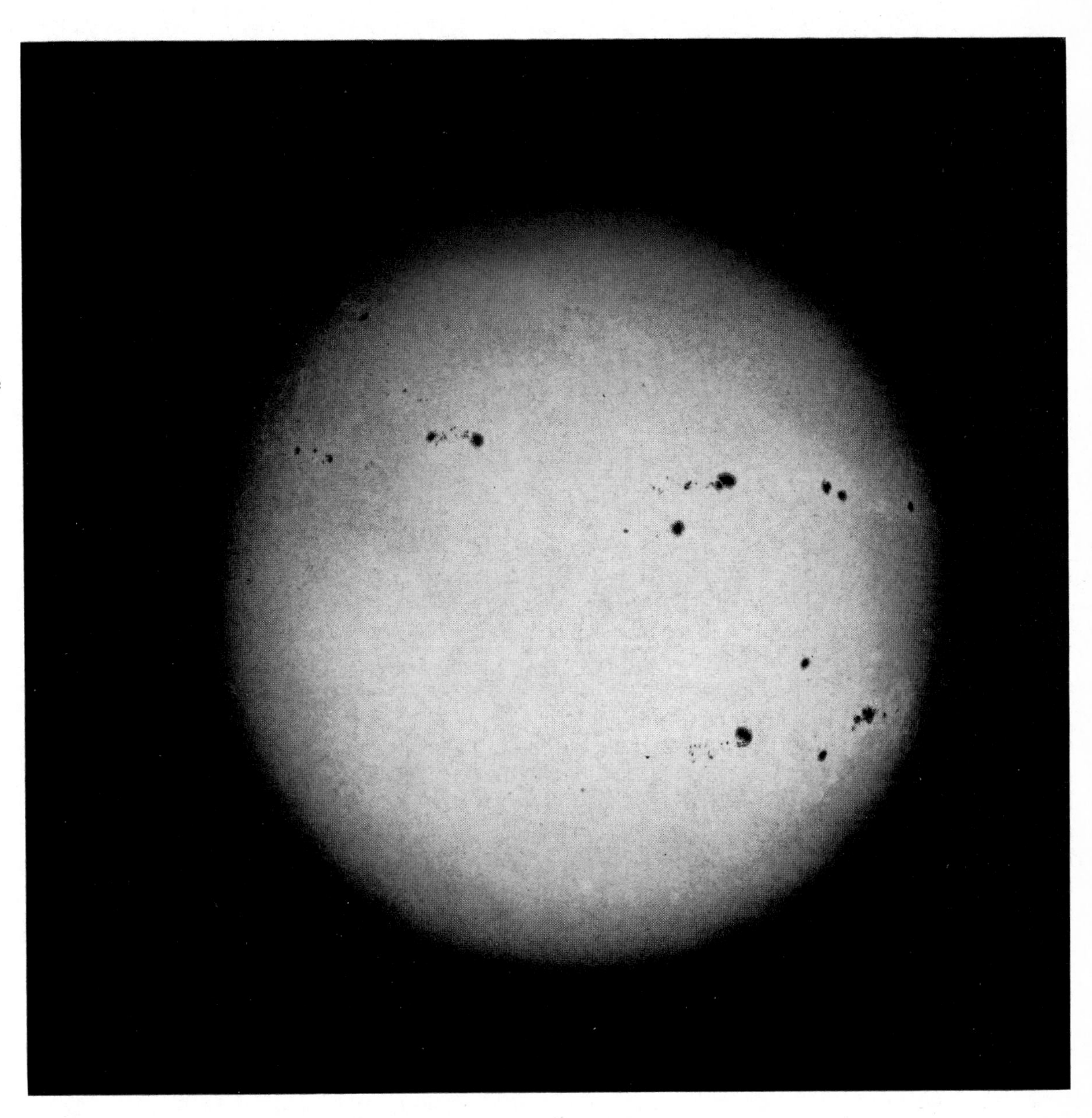

76 This general picture of the Sun shows, in addition to the darkening at the edge, a series of single spots and groups of spots on both sides of the solar equator. In some of the larger sunspots, the umbra, the dark central area, can be distinguished from the penumbra which is somewhat lighter. At various points near the edge of the Sun and in the vicinity of some of the spots, bright, network-like patterns known as solar faculae, can be recognized. These are clouds of gas in the higher layers of the Sun's atmosphere which are several hundred degrees hotter than their surroundings and the photosphere layers beneath them.

scopic investigations have shown that these are the upper part of a layer in which columns of hot gas rise from the interior of the Sun and, as they cool, sink back again. The photosphere is at a temperature of about 5,700 Kelvin which excludes the existence of solid or liquid substances.

Sunspots, flares and prominences

On very many days, the appearance of the photosphere described above is changed by disturbances of various kinds. The sunspots, which are relatively dark in appearance, are certainly the most familiar of these. Johannes Fabricius (1587—c. 1617), Galileo Galilei and Christoph Scheiner (1573—1650) were the first astronomers to observe the sunspots at the beginning of the Seventeenth Century with the aid of the newly-invented telescope. The small blemishes in the make-up of the Sun, which until then had been considered immaculate, caused considerable excitement and confusion at the time.

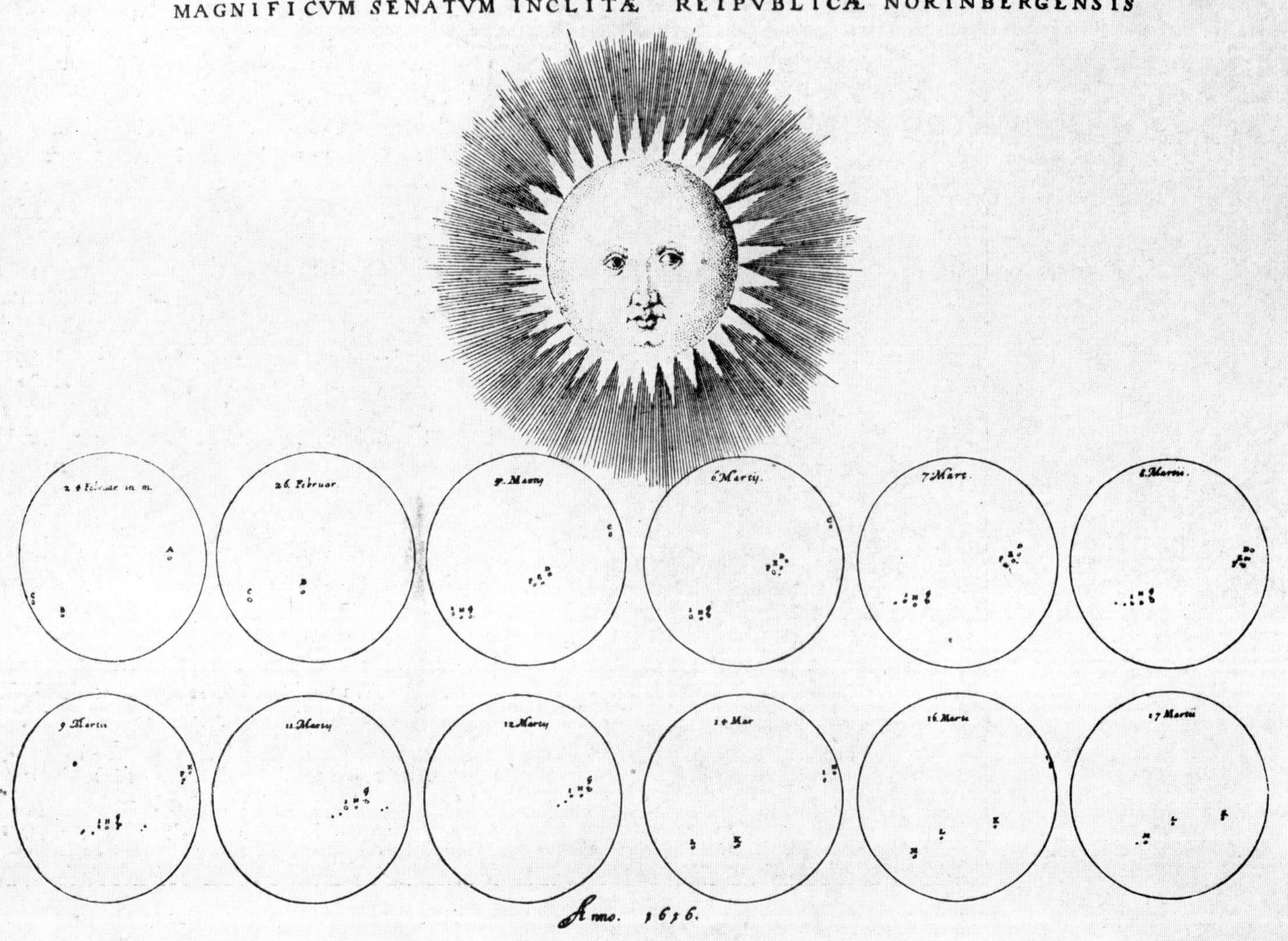

77 This old print is a compilation of systematic observations of sunspots carried out from February 24 to March 17, 1616. The movement of the spots across the Sun is due to its rotation.

78a, b The granulation and inner structure of a sunspot is particularly apparent from this photograph taken through a telescope carried by a balloon (a). The scale drawing of the Earth conveys an impression of the size of the granules and of the small sunspot. The life of a sunspot is a question of days or weeks. On the other hand the "life expectancy" of the individual granules is only a few minutes. It is a remarkable fact that the delicate radial structures in the sunspots were already noted a hundred years ago (b) (from a drawing by Secchi, 1872).

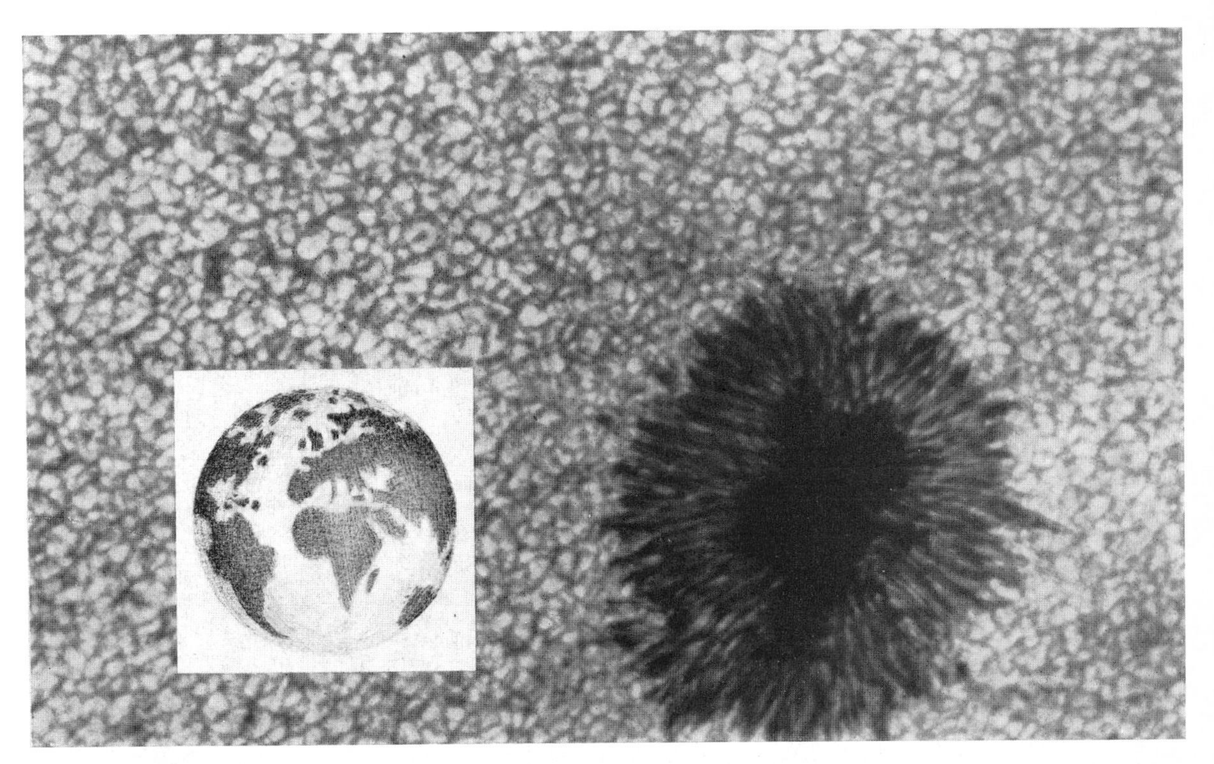

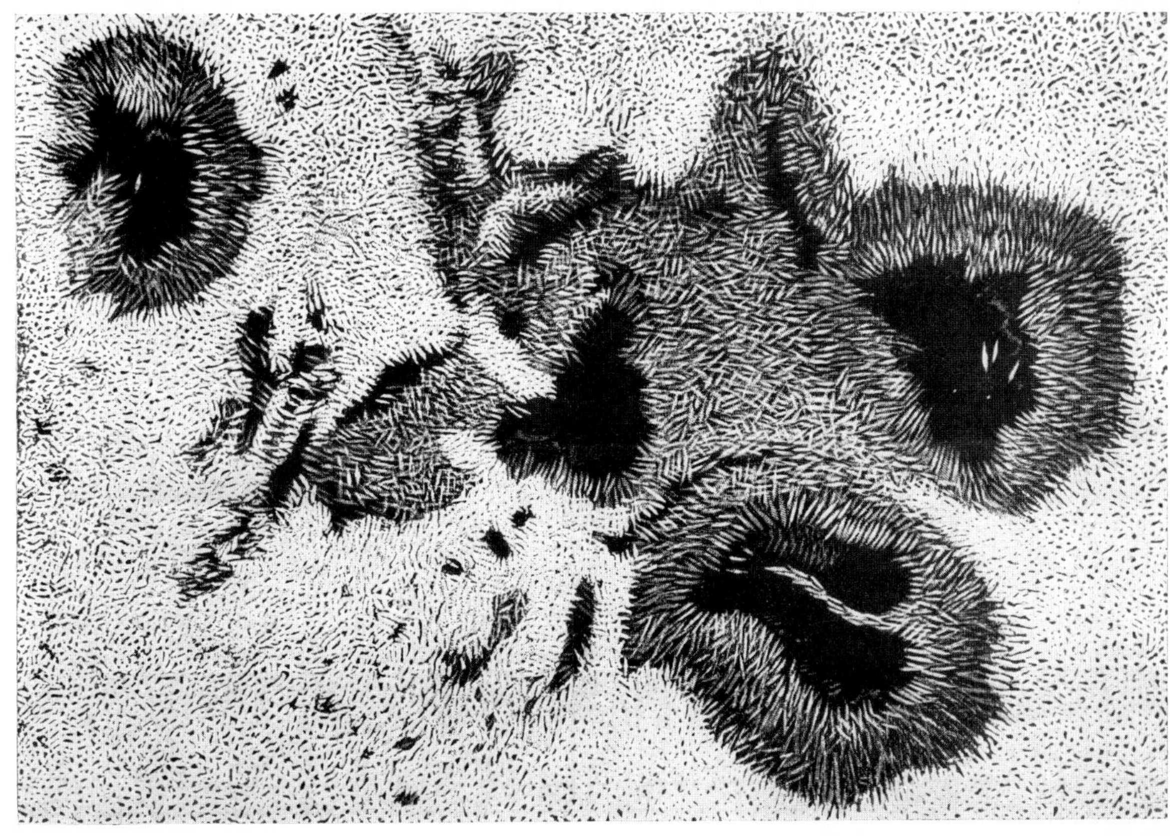

The size of the spots varies greatly, ranging from small points which can just be detected with a telescope to areas which are so large that they can be seen with the naked eye. In addition to single sunspots and pairs, whole groups of sunspots may appear, covering an appreciable part of the Sun's surface. Sunspots are relatively cool places in the photosphere, their temperature only being about 4,500 Kelvin. Localized magnetic fields of high intensity pass through the photosphere at these points.
It is astonishing that the temperature of the gas in the sunspots does not rise to that of the adjacent hotter areas. The answer to this riddle is probably that the magnetic fields of the sunspots indirectly act as thermal insulation. On photographs of pairs of sunspots taken through filters, the course of the magnetic lines of force can easily be identified by the particularly characteristic pattern of the clouds of gas at varying temperatures. Since each of the two sunspots forming a pair is of opposite polarity to the other, there is a great similarity between this and the patterns obtained when a horseshoe magnet is dipped into iron filings.
It often happens that a clear increase in brightness can be observed in the photosphere around the sunspots. Such areas are known as faculae. They consist of clouds of gas in a layer known as the chromosphere above the photosphere and are several hundred degrees hotter than the surrounding gas.
It was during total eclipses of the Sun, when the intense light of the Sun was obscured by the Moon, that the first observations were made of bright red arches and plumes projecting beyond the edge of the Sun and embedded in a pale light. These phenomena occur in a number of different shapes and are called prominences. Mostly they are long, laminar forms, usually rising like the arches of a bridge from the photosphere. They are found near sunspots and flares and sometimes remain in a quiescent state for months on end. This is ended by an active phase in which major changes in the shape of the prominence take place within a matter of hours. At this time, some prominences ascend to heights approximately equivalent to the diameter of the Sun, their form being substantially modified in the process.
With the appropriate observation methods and specially designed instruments, prominences can be observed at any time and even made visible in front of the Sun's disk. In addition to the relatively protracted disturbances of the photosphere already mentioned,

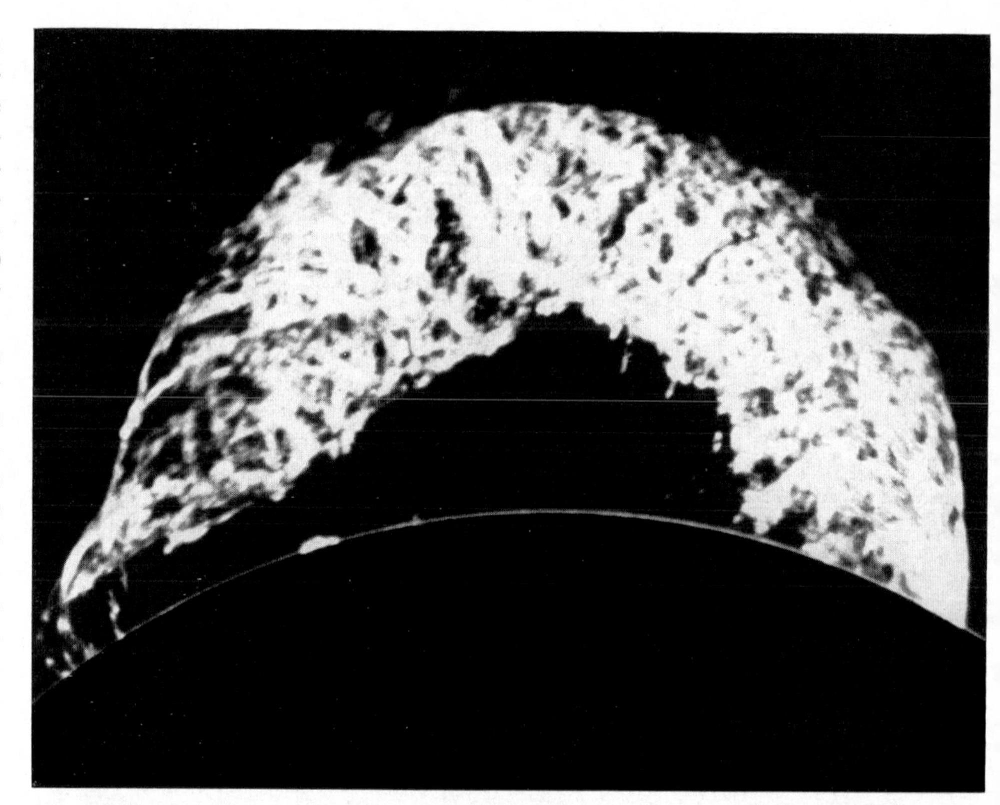

79a—c When the hot disk of the Sun is covered by a screen, one can always see prominences of a more or less bizarre shape rising above the edge of the Sun (a and b). Their sometimes rapid changes in shape can be easily followed. The frequency of the prominences is associated with the 11-year sunspot cycle. The flaming red colors of the prominences (c) are caused by the radiation of a special emission line of hydrogen.

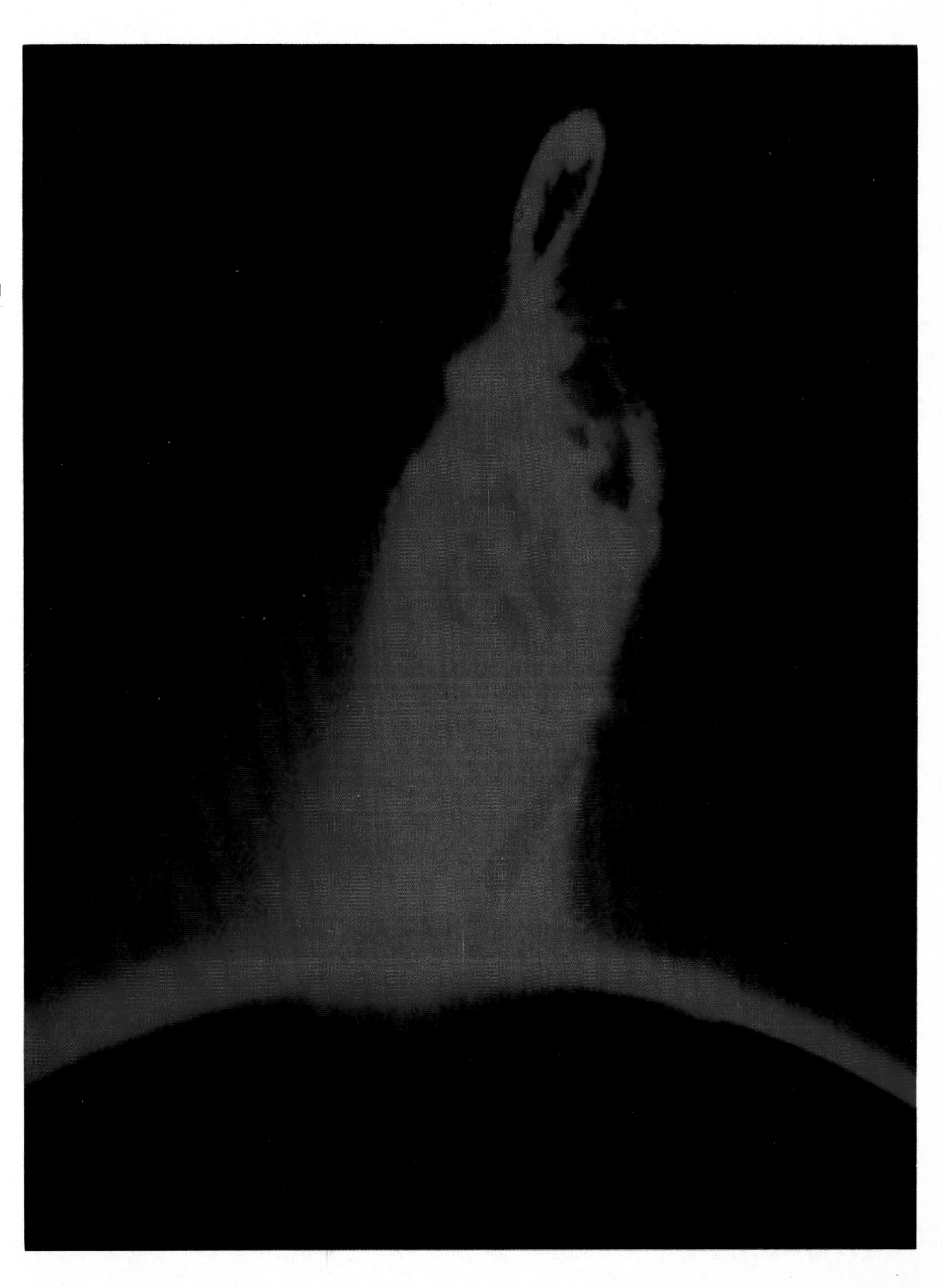

80 This composite photograph shows the structure and the extent of the corona and the surface of the Sun in the light of singly ionized calcium. There are bright centers of activity near the dark spots. The long dark forms are prominences projected on the Sun's disk.

short-lived increases in brightness, lasting only for minutes, occur in localized areas. In the course of an explosion of this kind on the Sun, there is an increased emission of ultraviolet radiation that is not visible to the human eye. It is considerably more than the normal level but is still not a great deal when compared with the total radiation of the Sun. The disturbance radiation and especially the constant ultraviolet radiation is absorbed by the upper layers of the terrestrial atmosphere so that a harmful influence on life on Earth need not be feared.

The solar "crown"

The outermost layer of the extended solar atmosphere is the corona. It was first noticed during total eclipses of the Sun when it was seen that the darkened Sun was enclosed in a silvery, crown-like pattern of rays. There is a gradual transition in the solar atmosphere starting at the corona and moving towards interplanetary Space. Compared with the density of the gas in the photosphere, the density in the corona is many magnitudes less, but exceptionally high temperatures are found there. This is why most of the electron shells are separated from the atoms. A gas of this kind, consisting of partially ionized atoms, ions and free electrons, is known as plasma.

The appearance of the corona is subject to chronological changes. Among its characteristic features are the bright streamers, forming an approximately radial pattern, at the pole areas of the Sun. These can always be observed, and the gas there marks the lines of force of the general magnetic field of the Sun, which is somewhat stronger than the magnetic field of the Earth.

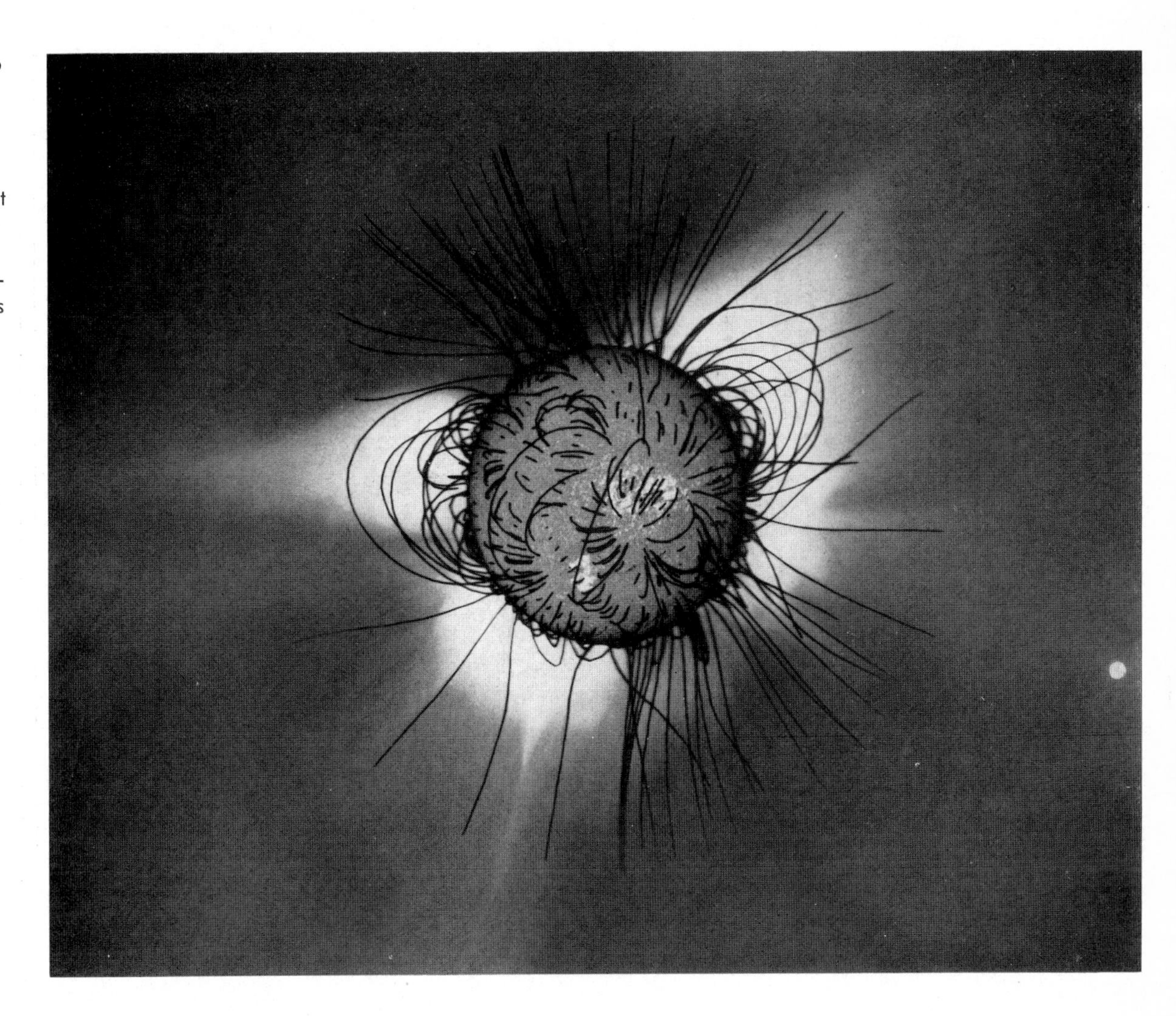

81 This illustration combines two photographs, one of them is a filter photograph of the Sun's surface, the other shows the corona. The centers of activity can be seen on the Sun's disk as bright areas. The dark lines are to demonstrate the course of the magnetic fields at the moment the photograph was taken. The Sun's axis of rotation is slightly inclined to the vertical, passing from the upper left to the lower right. In the vicinity of the rotational poles the lines of force of the large general magnetic field pass through the Sun's surface. The smaller arched lines of force are local and are partly associated with the visible centers of activity.

Radiation in the radio frequency region is also emitted from the active areas of the photosphere and the corona, but is of very low intensity in comparison with the optical radiation, being detected without difficulty only because the Sun is so close. In addition to slow fluctuations in the radio frequency radiation, radiation surges lasting only a few seconds are also known.

These readily apparent phenomena in the photosphere and the layers above it are major features of solar activity. As demonstrated by systematic observations over a period of many years, these phenomena increase at certain periods.

Rhythmic cycles and constant wind

In the middle of the Nineteenth Century it was noticed that the number of sunspots change in a rhythm of about eleven years. Analogous fluctuations in the other characteristic solar phenomena were subsequently shown to take place in the course of a cycle of 11 years. However, one now speaks of a 22-year cycle of solar activity since, as shown by the statistical evaluation of thousands of the short-lived sunspots, only after the completion of this period do the magnetic fields of the new sunspots display similar characteristics.

As far as it has been possible to explain these phenomena, which are intimately associated with each other, it is the interrelation between powerful local magnetic fields and streams of gas in the layers near the surface which is the actual motor controlling and maintaining solar activity. Its wide-ranging influence extends far beyond the orbit of the Earth up to the outer regions of our solar system.

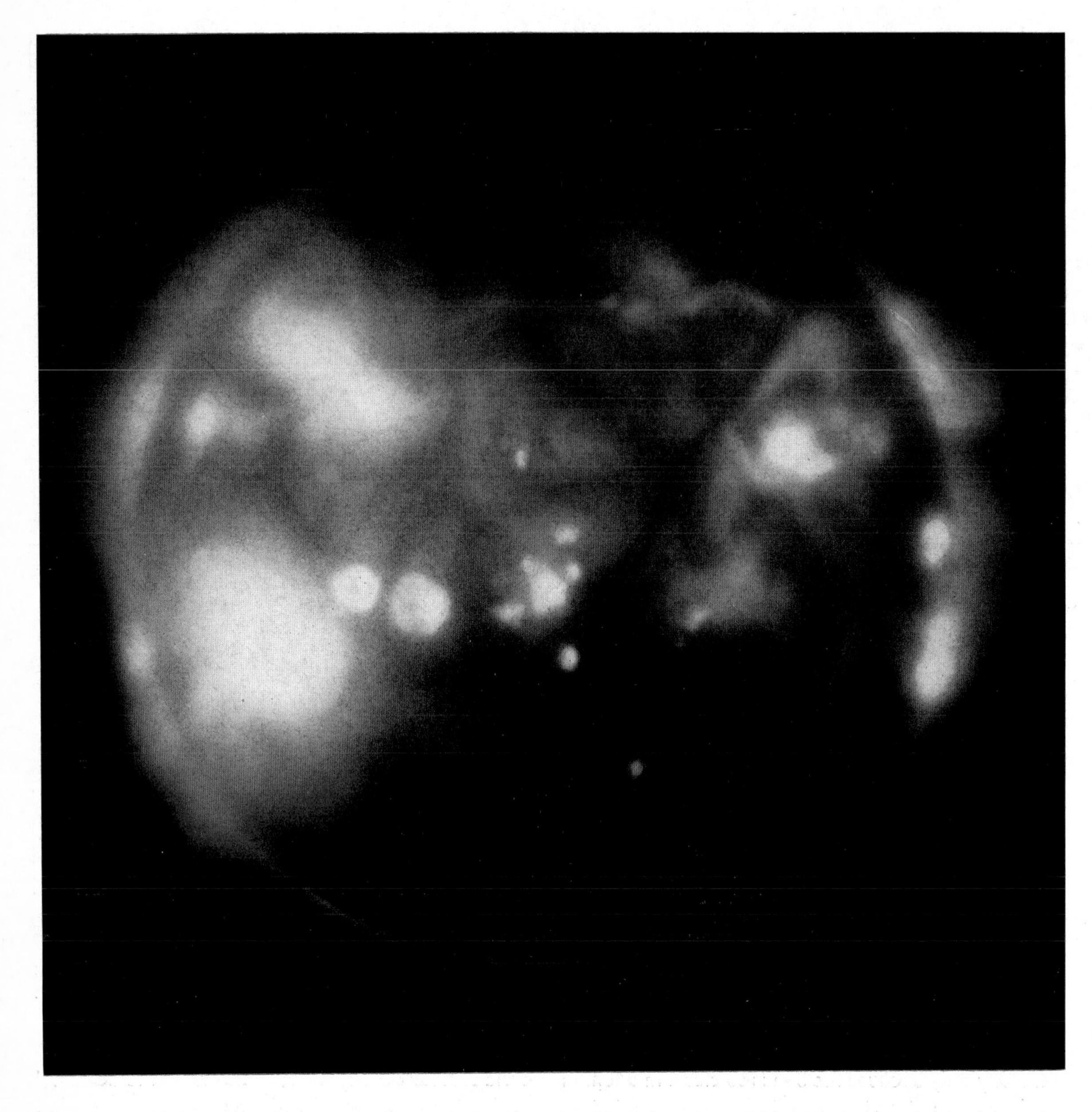

82 X-rays coming from the Sun can be recorded above the atmosphere of the Earth by instruments carried by a rocket or a terrestrial satellite. The picture shows several large bright areas, some of which extend beyond the edge of the Sun, thus increasing the three-dimensional impression. They mark the hot regions in the corona above the centers of activity. The thin bright ring around the Sun's disk indicates that X-ray radiation is also emitted outside the centers of activity, namely, from the undisturbed photosphere.

In addition to light, the Sun constantly emits in every direction a weak stream of electrons and atomic nuclei, this being known as the solar wind. Individual clouds of greater density and consisting of such particles and atomic nuclei, emitted by regions of activity, pass through it at a speed of several thousand kilometers per second.
When the Earth meets a plasma cloud of this kind, a number of interrelated effects may be observed. The most familiar of these are short-lived but powerful disturbances in the magnetic field of the Earth, terrestrial magnetic storms, radio interference in the shortwave area and the magnificent displays of the Northern and Southern Lights which can occasionally be seen even in the middle latitudes of the Earth. The Earth is shielded from the impinging particles by the atmosphere, its protective cloak.

The solar interior

Objective reasons prevent us from seeing into the interior of the Sun and the stars, but nothing stops us, in our imagination, from sending a probe equipped with all the necessary instruments into the stars. Its "instrumentation" consists of a number of important physical laws from which a computer calculates the "data" expected from the measurements taken along the course envisaged. The totality of the data thus obtained is designated as a solar model; the surface data from this is compared with information obtained from observations. If the differences which emerge are excessive, then the solar or stellar model, which is based on certain predetermined physical fundamentals, must be modified until a reasonable agreement is achieved between theory and practice. Calculations of this kind have produced the following approximate impression of the interior of the Sun. In a relatively small spherical region around the center is the atomic powerhouse of the Sun. The temperature and density of gas there are so high that a large number of atomic fusion processes can take place and can gradually lead to the production of a helium nucleus from four hydrogen atoms. In the course of this, a fraction of the mass of the atomic nuclei is converted into radiation energy. In the atomic jostling prevailing there, the radiation is passed on in the form of standardized "energy packets" by the atomic nuclei which, in rapid alternation, "swallow" them and "regurgitate" them at lightning

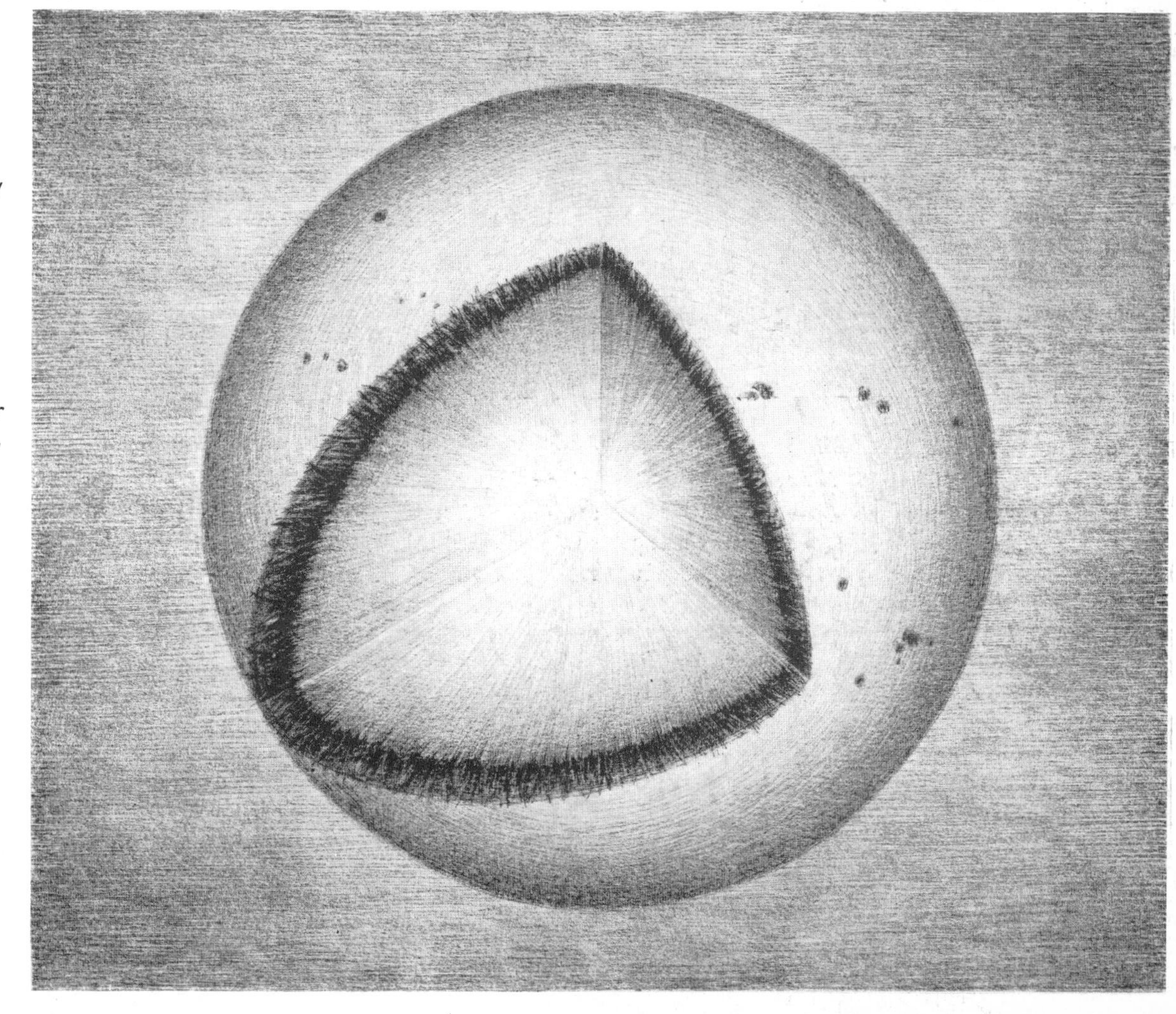

83 To demonstrate the physical conditions existing in the interior of the Sun, a section of the Sun's sphere has been removed. In the center of the Sun and in the region immediately around it, there are temperatures of approximately 15 million Kelvin and the highest gas density (about 100 g/cm^3). Here, the nuclear energy-releasing processes take place. In an area a little farther away from the central region, the nuclear "combustion zone," the temperatures and densities are much lower so that there are hardly any energy-releasing processes. In this area (the decreasing temperature is symbolized by an increasingly strong blue shading), the energy produced in the central region is merely transported in an outward direction. In a layer (the darkest in the illustration) reaching up to the surface, there are streams of gas rising and falling back again. This mechanism is called convection. Only the outermost layers of the Sun's surface may be observed visually.

speed, i.e., in less than a millionth of a second. It frequently happens that these energy packets break up into smaller units. In this way, they gradually move to the cooler edge of the actual combustion zone and enter an area in which temperature and density have fallen so much that hardly any energy-releasing processes still take place. Here the same transport mechanism continues to move energy in an outward direction, but at a certain depth below the surface an unexpected difficulty is encountered. The atoms there are not able to fully perform their task of "energy transporters" in the manner just described. However, they prevent the imminent pileup of energy in that whole clouds of them, charged with energy, rise to cooler regions and, after the radiation of the energy, fall back inwards again before repeating the cycle. This transport mechanism is called convection. The top part of this seething layer joins up with the photosphere, causing the surface granulation already mentioned.

The stars

Knowledge of the chemical composition of the celestial bodies is of fundamental importance for astronomy. The material nature of a whole series of different objects, including, for example, the Sun, a number of stars, clouds of gas in the space between them, the crust of the Earth, the surface of the Moon and meteorites, have been determined to a satisfactory degree of accuracy.

A common characteristic of cosmic matter in the gaseous state is the exceptionally high content of hydrogen, followed at a clear distance by helium. All the other chemical elements, even when taken together, account for only a few percent of the total mass of cosmic matter in the gaseous state. In contrast to this, the chemical composition of the solid bodies is characterized by a considerably smaller percentage of hydrogen and helium, which are even less common than the heavier elements. If, however, a comparison is made of the frequency distribution of

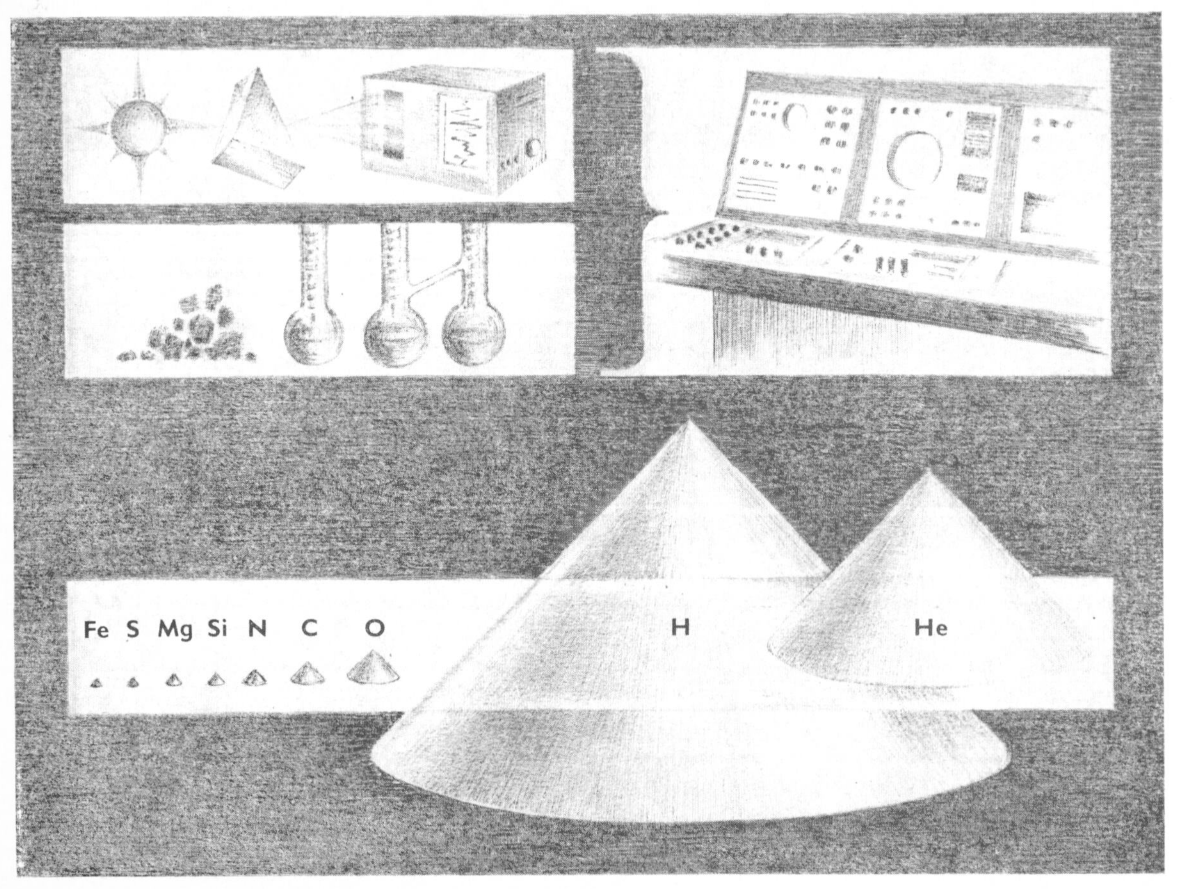

84 The average cosmic frequency distribution of the chemical elements is calculated from spectroscopic analyses of the light from stars and gaseous nebulae and from the chemical analysis of rocks. If the atoms of the chemical elements were to be heaped up according to their frequency, one would obtain, for the most common of them, the cones depicted in the illustration (hydrogen: H, helium: He, oxygen: O, carbon: C, nitrogen: N, silicon: Si, magnesium: Mg, sulphur: S, iron: Fe).

the heavier elements in gaseous and solid matter, a surprising level of agreement is found. Since most cosmic matter is in a gaseous state, hydrogen and helium are the most common of all chemical elements in Space.

Pursuing measurable quantities

It is very likely that the variety of the surface phenomena of the stars is similar to that of the Sun, and they probably have related forms of solar activity as well. Due to the great distances involved, however, it is not possible to detect any particular details on the surfaces of the stars. In contrast to the Sun, for the stars we have to be satisfied with an approximate physical description on the basis of observable parameters. These are the quantities that characterize the stars as a whole. The physical parameters can also be quoted for the Sun, of course.

The physical parameters of a star which are most useful for certain astrophysical investigations are its mass and its original chemical composition. Heinrich Vogt (1890—1968) and Henry Norris Russell (1877—1957) were the first astronomers to recognize the fundamental importance of these two measurable quantities for the structure of stars and summarized their findings in the uniqueness theorem of stellar structure named for them. This constitutes the theoretical foundation for the understanding of the entire development through which all stars pass.

Other important observable parameters of the stars include their diameter, surface temperature, brightness and rotational velocity. So that all the measurable quantities of the stars could be ascertained with the greatest accuracy possible, special methods had to be devised. Unfortunately, technical factors do not allow the required magnitudes to be determined for all stars. Thus, for instance, the direct measurement of stellar diameter is only possible in the case of a limited number of stars which are not so far away.

For other physical parameters, approximate numbers of the correct order of magnitude are the best that can be inferred. The lack of knowledge of the physical parameters of the stars, which in some respects is considerable, is compensated, however, by the wealth of observation material recorded for hundreds or thousands of other stars. The statistical processing of this material permits reasonably reliable conclusions to be drawn about the magnitudes of the quantities in question.

When comparing the measurable quantities of different stars, it is useful to take the equivalent figures of the Sun as reference values since a string of numbers does not always give a readily understandable idea of the differences involved between the stars, these being very great in many cases. As illustrated by the following data on the measurable quantities observed of the stars, when the totality of all these stars is considered, our Sun is only a small star of modest size in an army of similar celestial bodies.

If a statistical survey of the masses of the stars is made, it will be found that they range from 0.04 to almost 100 times the mass of the Sun. Most stars, however, fall within the much narrower range of 0.3 to 3 times the mass of the Sun. Considered numerically, the stars with a high mass are far less common than those with a low mass. This frequency distribution will be examined in more detail later when the theories on the formation of the stars are discussed.

Normally, the surface temperatures of most stars are in the range of 2,000 to 40,000 Kelvin, the stars with the greatest mass being the hottest. Some stars do not fall within this temperature range. The central stars of planetary nebulae have surface temperatures of up to 100,000 Kelvin, while on the other hand there are very cool objects, temperatures in some cases being less than 1,000 Kelvin. They are characterized chiefly by the emission of intensive infrared radiation. The nature of these infrared stars, as they are termed, is still not quite clear. It is conceivable that they are in the formation stage, i.e., extremely young contracting stars, but it is also possible that the infrared objects are stars with masses of less than 0.04 of that of the Sun. Low-mass stars of this kind cannot develop enough heat in their core to permit the release of energy on the basis of hydrogen "combustion." The consequence of this is that these stars, which are also known as "black dwarfs," slowly cool down and emit the low quantity of energy released mainly in the infrared range.

Giants and dwarfs

The diameters of the stars vary greatly. Thus, exceptionally large stars are known to exist which would have ample room not only for the Sun but also for the central parts of the solar system as well. On the other hand, there are also extremely small stars which, for historical reasons, are known as "white dwarfs." The size of these may be compared

85 From the occurrence and the thickness of the absorption lines in the spectra of the stars, conclusions can be drawn about the pressure, temperature and chemical composition of the stellar atmospheres. Depending on their temperature, one or other color component predominates, which is why the bright stars have a coloring that can even be identified with the naked eye. Starting at the top, the illustration shows the spectra of Spica, Procyon, Arcturus and Betelgeuse, the brightest stars in the constellations of Virgo, Canis Minor, Boötes and Orion.

86 About a thousand planetary nebulae are known to exist in our Milky Way system. They are made luminescent by the ultraviolet radiation of the very hot central star. The blue color of the central star in the Ring nebula of the constellation Lyra can be clearly seen. On an average, the mass of the luminescent gas is about 20 percent of the mass of the Sun. The diameter of the cloud of gas, which is at present about 5,000 million kilometers (3,000 million miles) in size, is expanding at a rate of about 30 km/s. It is thought that planetary nebulae are formed during the advanced stages of development of the stars and dissolve in the course of about 20,000 years.

with that of the Earth. However, since the equivalent of the mass of the Sun is accommodated in a relatively small volume, they are characterized by a fairly high density. On an average, one cubic centimeter of the hot, dense gas has a mass of 100,000 kilograms (100 metric tons). Thus, an amount of the white dwarf Sirius B the size of a pack of king-size cigarettes would weigh 10,000 metric tons or 22,000 English tons.

Considerably smaller than even the white dwarfs are the neutron stars, predicted in the 1930's. In the light of theoretical considerations, it is believed that they contain something like the quantity of gas in the Sun, concentrated in a sphere 10 to 15 kilometers (6 to 9 miles) in diameter. This implies a gas density of about a thousand million kilograms per cubic centimeter, a proposition which seems almost inconceivable to us. There is much evidence to indicate that the pulsars, discovered by chance a few years ago by radio astronomers, possess the predicted characteristics of the neutron stars.

The stars can be classified by their size as supergiants, giants, dwarfs and white dwarfs. According to this classification, the Sun is one of the dwarf stars. The necessity of distinguishing between giant stars and dwarfs was pointed out at the beginning of this century by the Danish astronomer Ejnar Hertzsprung (1873—1963) when he found out that there are stars which have the same surface temperature but different luminosity. Russell, the American astronomer already mentioned, subsequently compiled a diagram correlating the temperatures and luminosities found observationally for all the stars available, and he discovered that stars are actually found in only two regions. The Hertzsprung-Russell diagram, as it was later named in honor of the two famous astronomers, shows that Nature obviously favors certain combinations of temperature and luminosity.

This diagram is of inestimable value in the comparison of theoretical findings of stellar evolution with the corresponding data obtained by observation.
There is another relation between the luminosity and the mass of the stars. The observations in question are in agreement with theoretical conclusions and indicate that the luminosity increases approximately in proportion to the third power of the mass; that means that stars with high masses possess extremely high luminosities. Reference to the important mass-luminosity relation will be made later when the evolution of the stars is described.

As with the diameter, the luminosity of stars varies greatly. The supergiants have the highest luminosity and send out into Space ten thousand to a hundred thousand times the energy emitted by the Sun. The luminosity of the giants is, on average, one hundred times the luminosity of the Sun. The white dwarfs, on the other hand, are very sparing with the quantity of energy at their disposal and the energy they emit is equivalent to only about one ten-thousandth of that of the Sun.

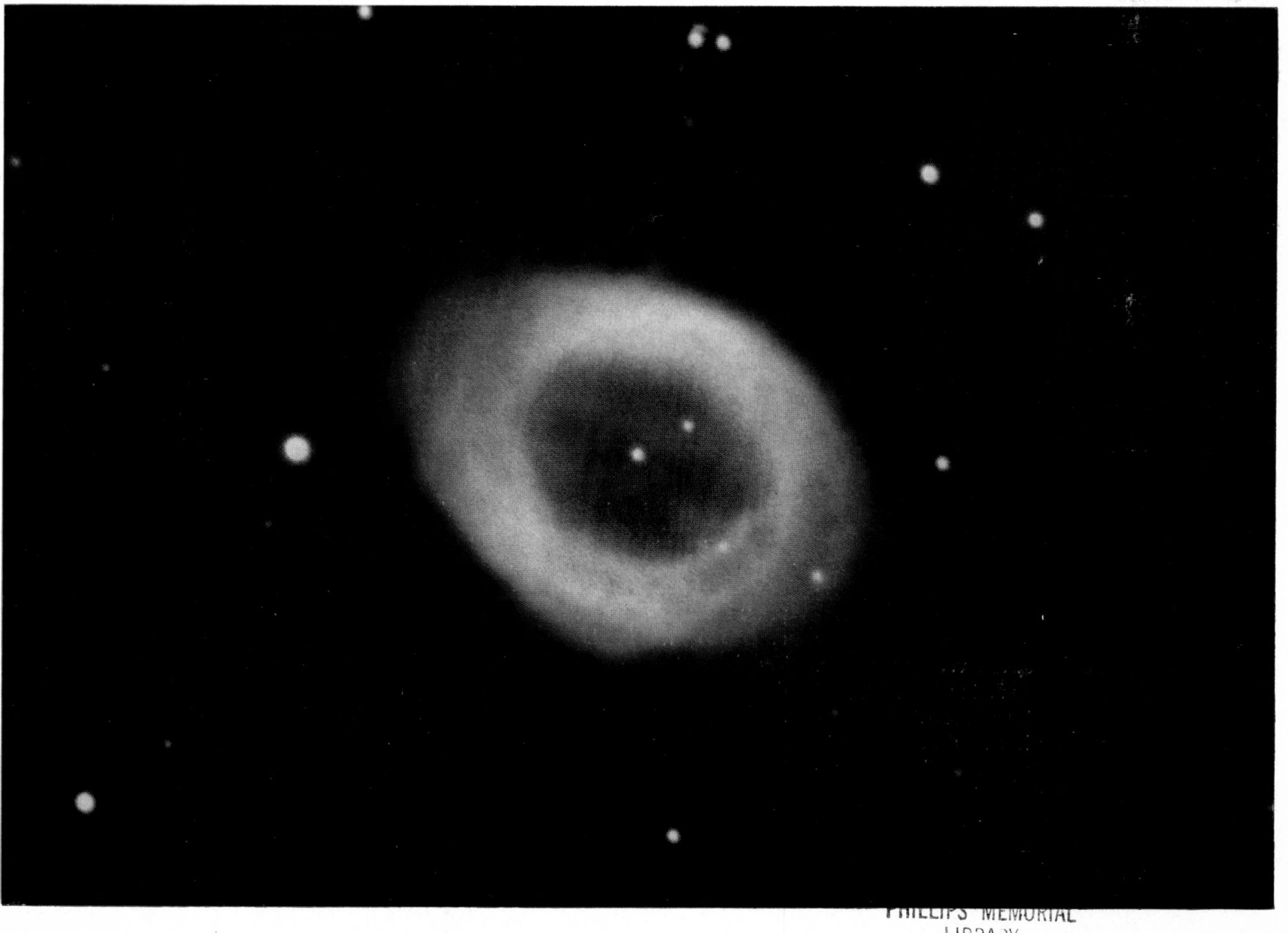

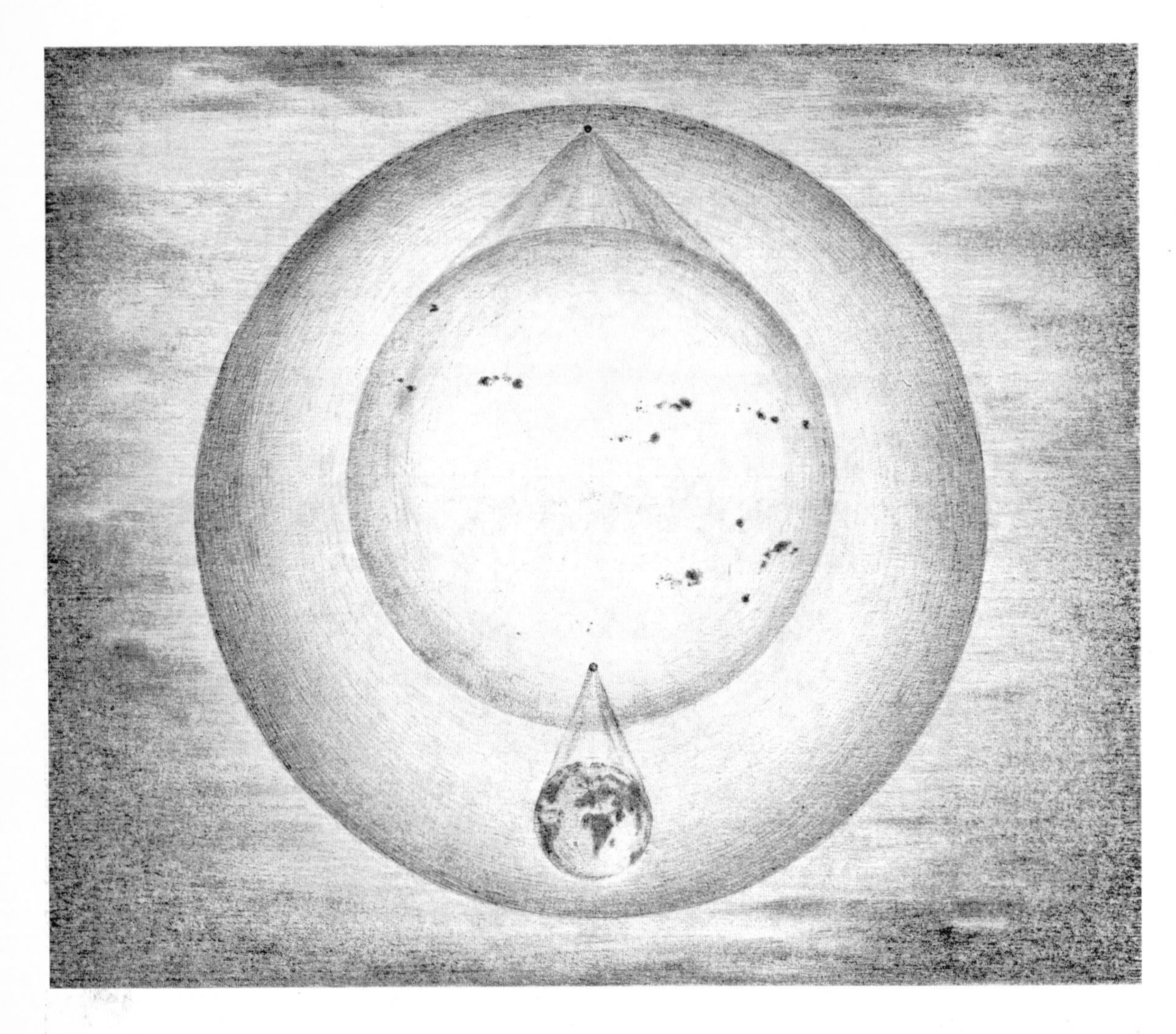

87 To understand the enormous differences between a giant star, the Sun and the Earth, it is almost necessary to use a magnifying glass to examine the accurate scale images of the smaller celestial bodies.

Stars with extravagances

Up to now, any reference here to the stars with extravagances has been discreetly avoided. These are a relatively large group of stars whose measurable quantities are subject to chronological fluctuations. A very wide range of different forms appear among the variable stars due to a variety of causes. Variations in luminosity were first noticed, revealed by fluctuations in the brightness of the light emitted by the star. The more or less marked changes in brightness can occur either periodically or irregularly. In some cases, the change in light is so small that it can only be identified with highly sensitive measuring instruments, but it can also assume gigantic proportions. Thus, in the course of a stellar "inferno," when a nova or supernova occurs, for instance, astonishing increases in brightness may be observed. With a supernova outburst, the brightness of the star in question increases by about a hundred million times within a few hours. It then takes several years before the enormous stream of energy again recedes to a low level.

The change in brightness of the true variable stars is due to physical processes in the external layers of the stars, and for the most part is associated with fluctuations in other measurable quantities. The nature and manner of the change in brightness and the generalization of the physical processes taking place in the stars concerned form the basis for the classification of the variable stars. So far, it has only been possible to explain the causes of variability of stars for a few types. This includes, for example, the

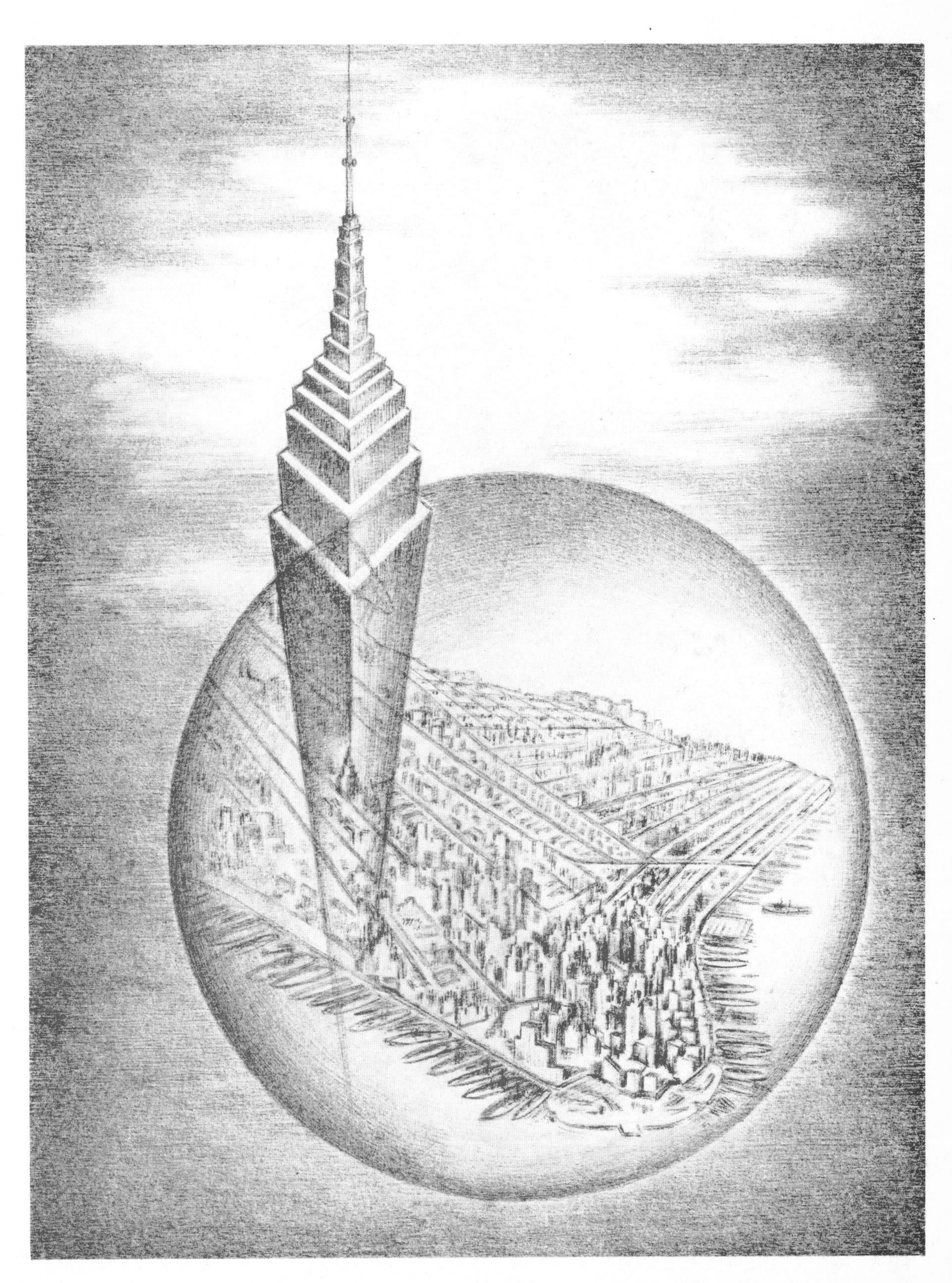

88 The diameter of neutron stars is about one-half the length of Manhattan, which is 13½ miles. It is believed that neutron stars possess a very strong magnetic field and rotate very rapidly. The emission of energy is said to be mainly along the lines of force. When the cone of light emitted from such stars happens to touch the Earth, it produces flashes of radiation similar to those observed in connection with pulsars.

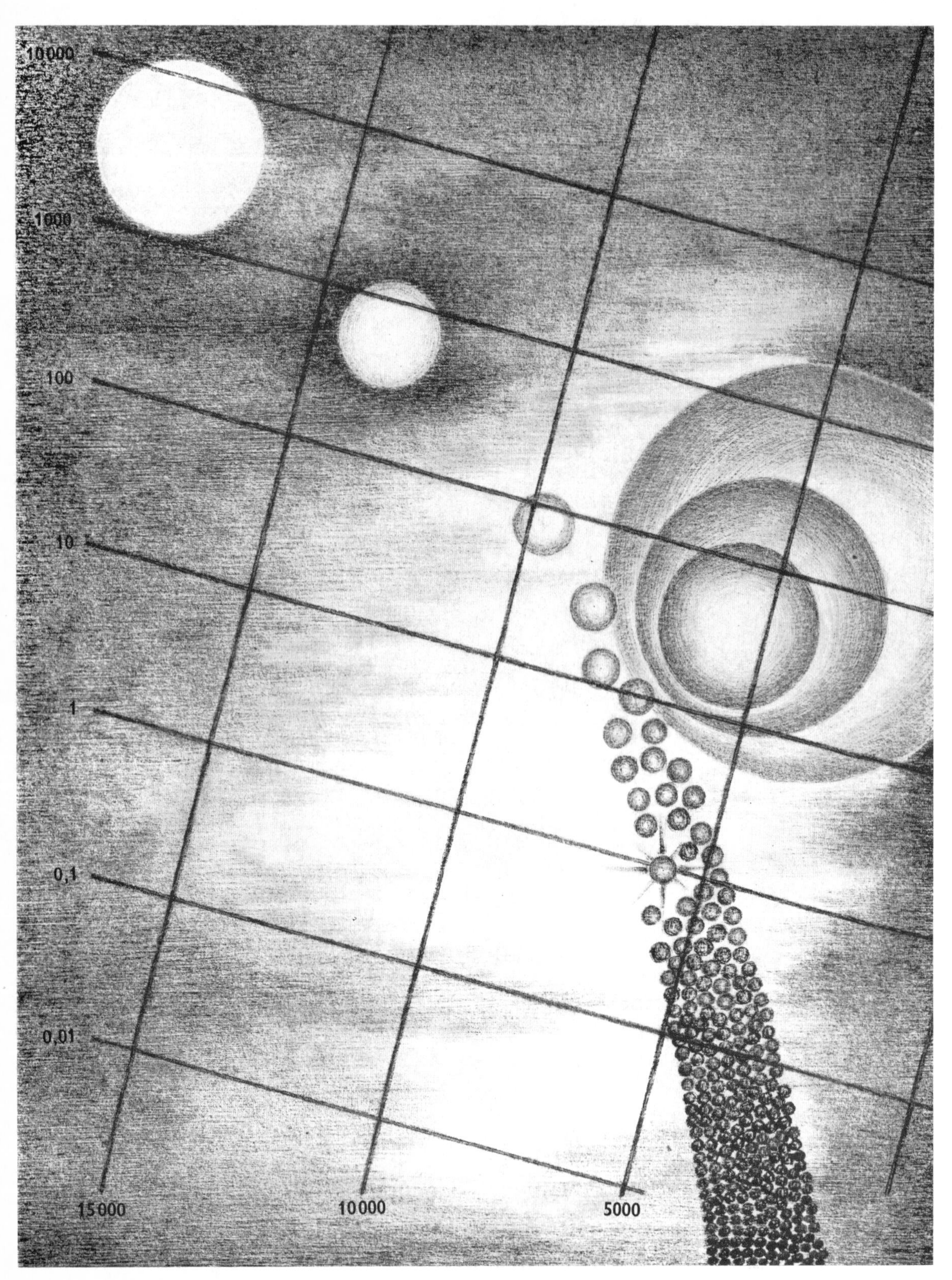

89 The Hertzsprung-Russell diagram records information about the surface temperatures and luminosities of the stars. The relative sizes and numbers correspond to the actual state of affairs. The area running diagonally across the diagram is called the main sequence because most of the stars are found there. They are termed dwarf or main-sequence stars. Above the main sequence, there are the giant stars. On account of their low luminosity, the white dwarfs would have to be recorded far below the main sequence. The position of the Sun in the diagram is indicated by rays. The stream of energy—its luminosity—constantly emitted by the Sun is used as a basis of comparison for the luminosities of the other stars.

group of pulsating stars. These stars alternately contract and expand according to a more or less precise timetable. It could be said that they "breathe," the frequency of breathing being dependent on the luminosity. Hence there are regular relations between luminosity and pulsation period for variables of the Delta-Cephei and W-Virginis types. They are of great importance for determining the distance of faraway objects such as the stellar systems.
Our present knowledge of the causes of variability indicates that many of its forms are obviously linked with quite definite stages of stellar development. Some kinds of variability may be explained as "teething troubles" of the stars, the consequences of which are of little significance for the further development of the star. Other forms can be regarded, with a certain justification, as typical "ageing phenomena" since they occur in the later stages of stellar development. In this connection, the investigation of the variable stars is throwing light on certain aspects of stellar development.

The binary stars

Like many other stars, our Sun is a solitary individual, separated by vast distances from even its closest neighbors. On the other hand, there are also the pairs of stars or double stars, representing an interesting group of cosmic objects. The relatively close components of a double star are linked by gravitation and consequently move around a common center of gravity. The duration of an orbit is determined by the distance separating the centers of the two component parts. When these are very close to each other, an entire orbit is completed in only a few days. Other double star components have such large orbits that a single circuit takes hundreds of years to complete. It seems that when Nature created the double star systems it wanted to demonstrate the contrariness of its moods.
A pair of stars, which is as familiar as it is unequal, is comprised of the brightest star in our sky, Sirius, and its companion. While Sirius is more than twice the size of the Sun, its companion is an insignificant white dwarf, comparable with the Earth in diameter. Other double star systems also unite stars of varying size and temperature. If one of them should be orbited by an inhabited planet, the people on the latter will certainly be able to observe fantastic lighting effects when the countryside is illuminated by both suns simultaneously in different colors.
The orbital position of some double stars is such that the orbital motion of the two components happens to intersect the line of sight to the Earth. As a result, the stars regularly and alternately assume a position in front of each other, this being revealed by a characteristic change in light.
A particularly interesting system of this kind is Zeta in the constellation of Auriga. The extensive atmosphere of the giant star is so thin that its smaller but hotter and consequently brighter companion, which periodically moves behind it, is still able to shine through the outer layers. For astronomers, this is an almost unique opportunity, during the cosmic transillumination of a star, to study the outer layers of its atmosphere.
For the unconditional determination of the masses of stars with the aid of the law of gravitation, the knowledge of exact double star orbits is extremely important. Unfortunately, the number of orbits accurately known is still very small, so that reliable data on their mass is available only for a few stars.

Star clusters

In addition to the double stars, systems are also known in which three, four or more stars are linked with each other by gravitation. Among the stars, which are generally distributed in a remarkably irregular fashion, there are the star clusters. In number these vary from a few dozen to a few million stars, separated from each other by a distance of "less" than ten million million kilometers or about one light-year. In passing, it may be noted that the stars of the solar environment are on the average four light-years away from each other. From the viewpoint of the history of their development, all the stars belonging to a loosely associated open star cluster form one unit, since they were formed at about the same time. With an appropriate observational technique, a distinction can be drawn between star clusters of differing age. The youngest of them were formed only a few million years ago, whereas the oldest star clusters have already existed for several thousand million years. Important empirical information for theoretical calculations of stellar development is obtained from investigations of open star clusters.
A much more compact form of star cluster is the globular cluster. As implied by their name, they have a largely spherical appearance. The number of stars in a globular cluster varies, but can amount to

90 Sirius and its companion, a white dwarf, move around their common center of gravity on orbits which are comparable to the planetary orbits in the solar system. On the photograph (inserted rectangle), the white dwarf, which is only the size of the Earth, is the bright point to the upper left of Sirius. Its hexagonal shape is due entirely to the exposure technique employed.

hundreds of thousands and even millions. Within a globular cluster the average distance between stars is only ten thousand times the distance of the Earth from the Sun. The globular clusters are among the oldest cosmic objects known and were probably formed about ten thousand million years ago.
An attractive idea is to imagine the Earth as the planet of a star in a great globular cluster. The sky which we would see would be fairly uniformly covered by tens of thousands of stars of the brightness of Sirius. The total light emitted by them would be so intense that the terrestrial atmosphere at night would be characterized by a perpetual twilight.

Luminous interstellar clouds

Already about 200 years ago, observations with telescopes had revealed a whole series of pale spots scattered throughout the sky and emitting a dull light. They were called nebulae. At that time men could only speculate about their nature, but since then systematic investigations have resulted in clear physical hypotheses. These indicate that the luminous nebulae include objects which are totally different from each other in composition and origin.
The following remarks apply only to those nebulae within the Milky Way. These are extensive clouds of gas and concentrations of microscopically small solid particles of dust, located in interstellar Space, the region between the stars. Even though the gas and the dust particles between the stars are extremely finely dispersed by terrestrial standards and represent only a tiny fraction of the mass of the stars, this interstellar matter is nevertheless of great astronomical significance. Among other things, interstellar matter is the "building material" that is utilized when favorable circumstances permit the formation of stars.
In the vicinity of luminous nebulae there is always at least one star of great luminosity and it is this which is responsible for the luminous appearance of the nebulae. Expressed more precisely, it is the ultraviolet radiation of the high-luminosity star which causes the atoms in the interstellar cloud to emit light. Without going into further details of the physical processes involved, the action of the hot stars may be compared with that of streetlights which, under foggy conditions, illuminate the immediate area around them. In general, the interstellar clouds of gas extend further than the area illuminated by the hot stars. If a star is situated in the vicinity of a cloud of dust particles or is even embedded in it, a fraction of the radiation from the star striking the tiny solid particles is reflected by them in all directions and makes it possible for all of them to be seen. Figuratively speaking, the dust particles may be imagined as tiny mirrors reflecting the light of the star in all directions like a gigantic crystal chandelier. The characteristics and significance of interstellar dust are discussed in more detail in the remarks on condensed matter.
The chemical composition of interstellar gas is roughly equivalent to the average cosmic frequency distribution of the elements. On the other hand, the observations made do not yet allow any clear-cut conclusions to be drawn about the material nature of the dust particles, but there are indications now that the dust consists mainly of silicates.
Pure nebulae of either gas or dust are seldom found, most of the interstellar matter being a mixture of the two. In the constellation of Orion, a nebula may be observed which is of particular astronomical interest. In all probability, it is the best known gas-and-dust nebula. In addition, it contains a fairly large number of stars, which belong to three different generations. It is estimated that the most senior of these is five million years old. The youngest stars, however, were formed only about 15,000 years ago and caused the nebula, which is even visible to the naked eye, to become incandescent. It seems that the process of the formation of stars is still continuing in the Orion nebula if the explanation of the infrared objects found there as extremely young stars is correct.
From the discussion of the observational material, it is apparent that the spatial proximity of luminous nebulae and stars of high luminosity is not accidental but expresses a genetic relationship. On the other hand, as already mentioned, the existence of interstellar matter is not restricted solely to the place where it is made incandescent by stars. Photographs of bright gas nebulae, for example, provided evidence of the presence of dark concentrations of dust. These bright gas nebulae often form a background against which the clouds of dust are seen as black forms. In other cases, the concentrations of interstellar dust have been identified by their light-absorbing effect, when they give the impression that there are no stars in certain parts of the sky.
As long ago as the beginning of the present century, the attention of astronomers was drawn by chance to the nonluminous interstellar gas. In spectroscopic

91 a—c A well-known open star cluster is the group known as the Pleiades, whose stars came into existence several million years

ago. It is estimated that this star cluster, which is about 400 light-years away and 30 light-years in size, consists of 300 to 500 stars of which the eight brightest may be seen with the naked eye when conditions are most favorable (a). Photographs taken with long exposure periods show that the bright stars are enclosed in bright nebular mist (b). This is made up of interstellar dust which reflects the radiation of the brighter stars. The delicate structures of nebulosity can only be seen on photographs taken with fairly large telescopes. Galileo produced two drawings of the Pleiades in 1610. One of them was based on observations made with the naked eye (c, the small drawing on the upper left) while the other incorporated what he was able to see through a telescope (c, the central portion). However, since the telescope could not be guided, the positions of the stars given are not accurate.

92 The M 13 globular star cluster is in the constellation Hercules. In all, there are about a hundred globular star clusters in the Milky Way system and they are distributed around it in a large spherical area. The globular star clusters have an average diameter of a hundred light-years and they are probably the oldest objects in the Milky Way system. They have also been found in other galaxies.

93a—d A gas-and-dust nebula of very great cosmogonic interest is the Orion nebula (a), which is about 1,500 light-years away from us. Its luminescence is caused by four very hot and extremely young stars, the stars of the Trapezium as they are known. The Trapezium stars can only be seen on photographs taken with a very short exposure time (b). The outer areas of the cloud complex, which is some ten to fifteen light-years in size, can easily be seen on the longer exposures. It is suspected that stars are still being formed in some parts of the Orion nebula. Furthermore, radio techniques have shown that there are small regions in which hydrogen, and different types of molecules such as hydroxyl, water, carbon monoxide, carbon monosulphide, hydrocyanic acid, formaldehyde and methyl alcohol occur in the gas state in high concentrations. The drawing (c) of the Orion nebula made by Christian Huygens in 1656 includes only its brightest parts. That by Charles Messier in 1771 (d) shows considerably more detail and the Trapezium stars can clearly be identified on it.

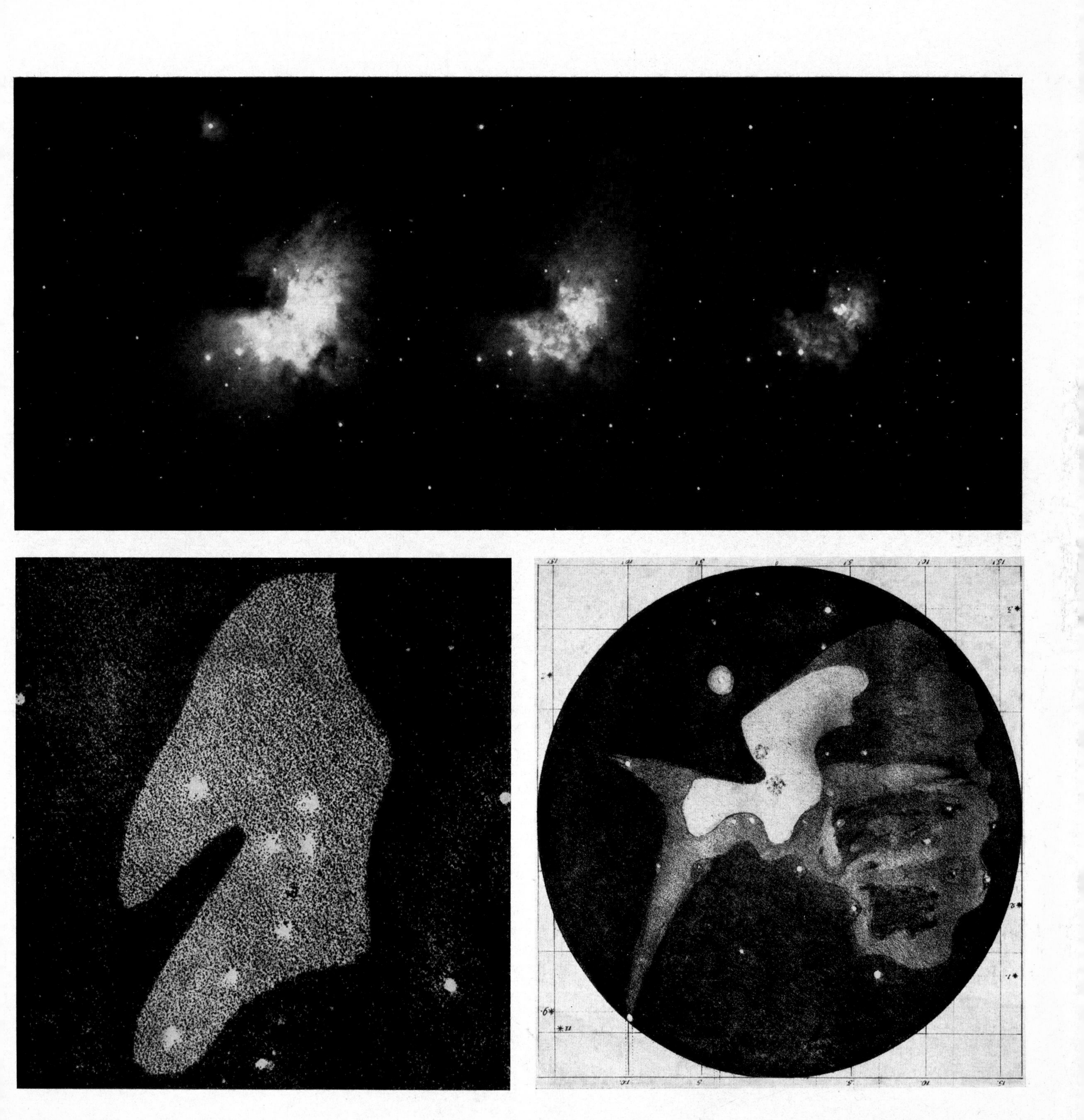

94a, b In the constellation Sagittarius there are the Lagoon nebula M 8 (a) and the Omega nebula (b).

The red light they emit comes from the radiation of luminescent interstellar hydrogen. Both these nebulae are about 30 light-years in diameter and have a total mass of about 1,000 times that of the Sun. The Lagoon nebula is about 5,000 light-years away from us. Bizarre scraps of dark matter in the dust form are projected on the luminescent gas of this nebula. The Omega nebula is just under 6,000 light-years away.

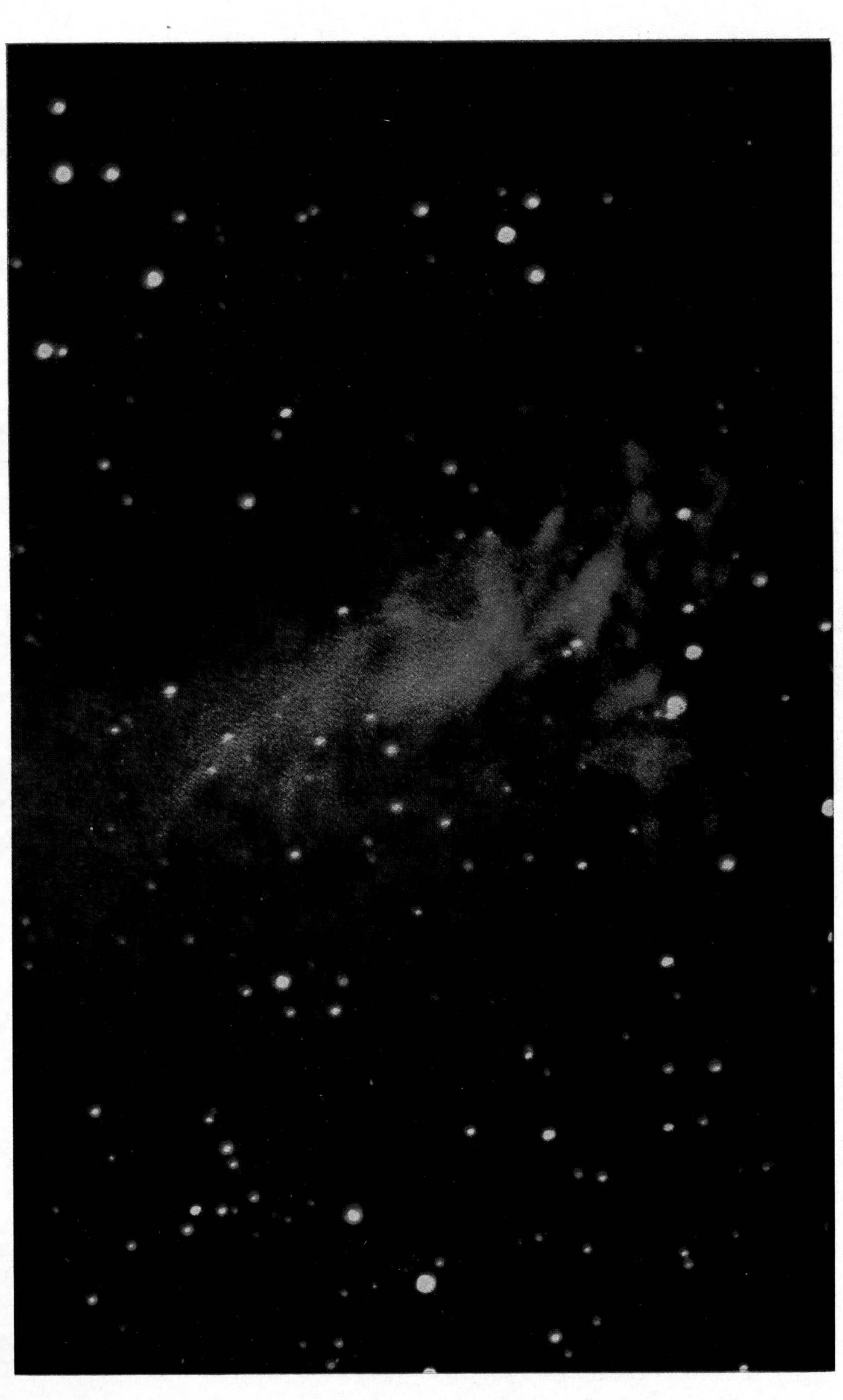

investigations of the orbital motion of extremely close double stars it was noticed that although the vast majority of the stellar absorption lines were blurred and pale there were also some very narrow ones among them. In a very perceptive examination of this phenomenon, Johannes Hartmann (1865—1936) showed that their cause was not to be found in the stellar atmospheres but in clouds in interstellar Space. Since then, stars have been discovered in whose spectra the interstellar lines are multiply split up. From this we can conclude that between us and the stars in question there are a number of interstellar clouds, moving at different speeds along the line of sight. About 20 light-years is estimated as the average extent of the clouds. The density of the gas within the clouds is very low and represents a better vacuum than it is possible to obtain in our laboratories at the present time. If we were to examine large quantities of this gas, we would find an average of about 10 hydrogen atoms in every cubic centimeter of gas, in some cases 100 atoms and very occasionally 1,000 atoms. Despite the low density of the gas, the enormous extent of the clouds means that their total mass can be more than a hundred times that of the Sun.

The average density of interstellar dust is a hundred times lower than even that of the gas. This means that in an area with an edge length of 100 meters there are only one or two microscopically small particles of dust.

Radio astronomy is playing an important part in the investigation of interstellar gas. Neutral hydrogen atoms emit radiation with a wavelength of about 21 centimeters, and this characteristic has been utilized by radio astronomy. Since, in contrast to stellar radiation, it is not absorbed by the interstellar dust, the 21-centimeter radiation can be picked up even at very great distances. From this, a picture may be gained of the large-scale distribution of interstellar hydrogen and of its movement in Space.

Many of the cloud conglomerations consisting of gas and dust contain cloud-like formations which are fairly dense and of strikingly small size. They are the source of intensive radio radiation on various wavelengths. As far as it has been possible to decode these radio messages reaching us, it seems that they originate from a whole series of different organic molecules.

95a, b The profusion of stars in the area of the Milky Way around Gamma in the constellation Cygnus is only really apparent on long-exposure photographs (a). The inserted rectangle marks that part of the field which was photographed with the biggest Schmidt reflector in the world. It shows a wealth of minute detail (b). Dark clouds of dust, some of which are very large, can be seen against the luminescent gas nebulae. A comparison of these two pictures is striking evidence of the excellent picture quality of the Schmidt reflector, the other photograph having been taken with a lens telescope.

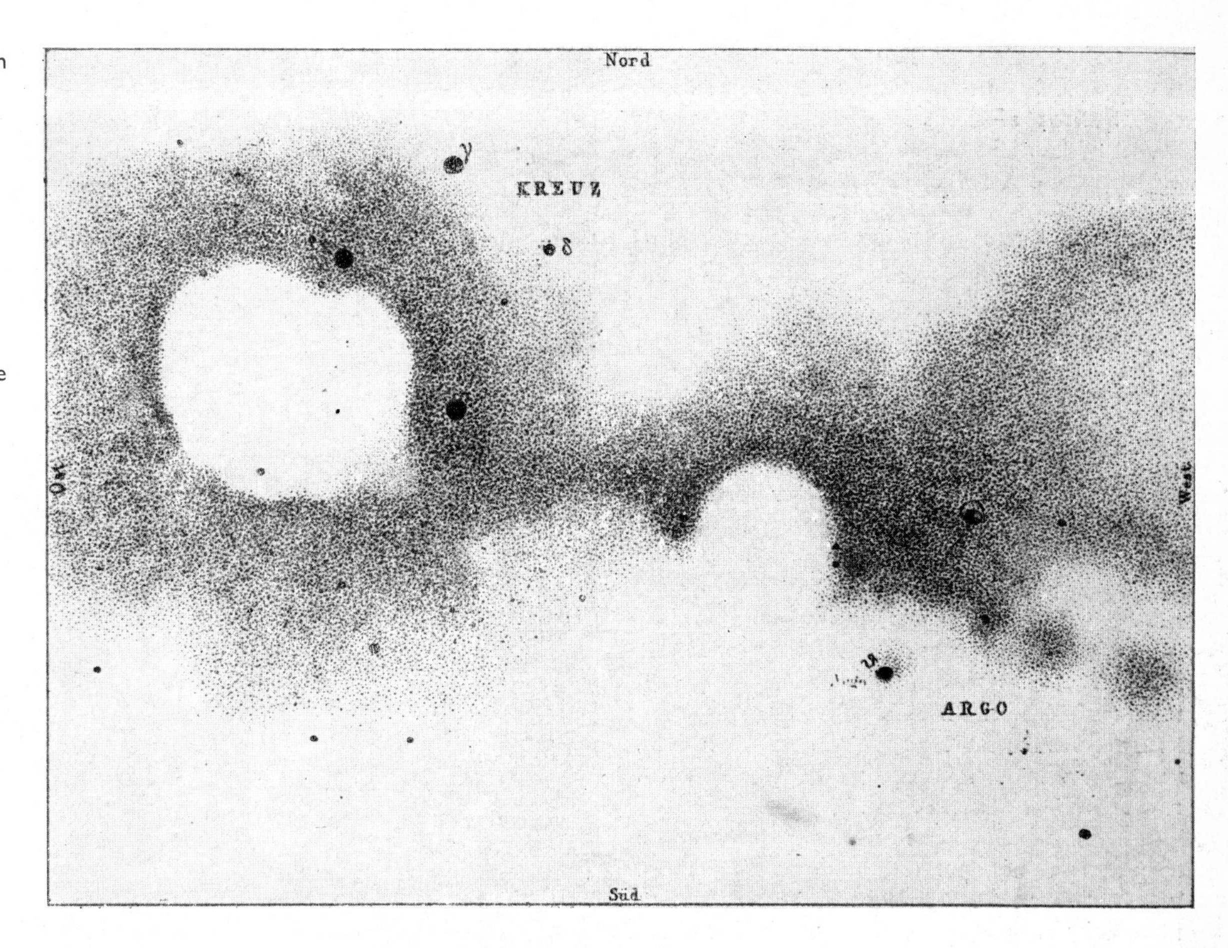

96 a, b In that part of the southern Milky Way which is near the Southern Cross and Centaurus constellations, a dark area known as the Coal Sack catches the eye (a). This is a dense cloud of interstellar dust that absorbs the light of the stars behind it. The two bright stars on the left of the picture are Alpha and Beta in the constellation Centaurus. On the upper right of the Coal Sack, there are the four bright stars which form the well-known constellation of the Southern Cross. The sketch (b) by Sir John Herschel is based on visual observations of this part of the sky (Kreuz = Crux or Southern Cross).

The vast structure of the Milky Way

So far, a general description has been given of objects which are frequently found in Space and of their spatial location with reference to their surroundings. The next question to be examined is whether the stars and the nebulae are infinitely distributed throughout Space or, in their totality, are part of a higher organizational unit. It was precisely this problem which attracted the attention of William Herschel more than 150 years ago. With the aid of the extensive observations which he systematically carried out, he came to the correct conclusion that the stars are arranged in a spatially limited and flattened system. In the evaluation of Herschel's work on this subject, however, it should not be forgotten that he could only see a small part of the complete system now known as the Milky Way system, galaxy or stellar system. In the meantime, reliable data has been collected which indicates that the vast structure of the Milky Way may be compared with a discus. Unfortunately, due to the position of the Sun within the galaxy, we cannot have a bird's-eye view of the entire system. Observations have shown, however, that the edge of the Milky Way which can be easily identified with the naked eye marks all those directions in Space in which the galactic disk extends. Naturally enough, there is an accumulation of stars in those directions and, in their totality, they appear to us to blend together and form a diffusely luminescent whole.

Before continuing with the examination of the structural features of our stellar system, a few remarks must be made about the nature of the spiral nebulae. For a long time, astronomers were unable to agree about them. Some considered that they were remote "island universes," while others believed them to be gigantic clouds of gas. It was only at the beginning of the 1920's that the American Edwin Hubble (1889—1953) solved the riddle of the spiral nebulae. From photographs he was able to identify some variable stars and further investigations proved that the spiral nebulae and the elliptic nebulae are indeed vast concentrations of stars. In addition, the specific characteristics of the variable stars found in the stellar systems enabled accurate data on distances to be obtained. Thus, for instance, the nearest major stellar system, the familiar Andromeda nebula and its two small companion galaxies, is about 2.2 million light-years away from us.

Stimulated by the great variety of the distant galaxies, Hubble's important discovery was soon followed by a systematic and successful search for the corresponding structural features in our own Milky Way system. From this, we now know that it is a spiral galaxy and that the Sun is located at the inner edge of one of the spiral arms, far away from the center of the galaxy.

The presence of spiral arms in the more distant environment of the Sun was demonstrated with the aid of the brightest stars and the young open star clusters. These objects of greatest absolute brightness make the spiral structure visually apparent. The distribution of the cooler and fainter stars, however, is smoother. These stars are situated in the spiral arms as well as between them.

Optical methods cannot be used for the observation of the more remote parts of the arms of the spiral since the clouds of dust concentrated in some regions of the Milky Way completely obscure the light from the stars behind them. The power of penetration of the radio telescope, however, is not affected by interstellar dust, and it has consequently proved possible to determine the large-area distribution of hydrogen for almost all the parts of the Milky Way system. The arrangement of the clouds of hydrogen in a flat layer is only disturbed to a major extent in the outer areas of the Milky Way. The disk form there is bent like the brim of a hat.

It is a pity that the central region of our galaxy is hidden by thick clouds of interstellar dust and thus cannot be observed visually. Nevertheless, it has been pinpointed with infrared and radio observations. They have not only revealed that the central region is in the direction of the constellation of Sagittarius but have also brought some very important details to light. Thus the entire central region covers an area of about 1,000 light-years in diameter while its nucleus of about 30 light-years is remarkably small.

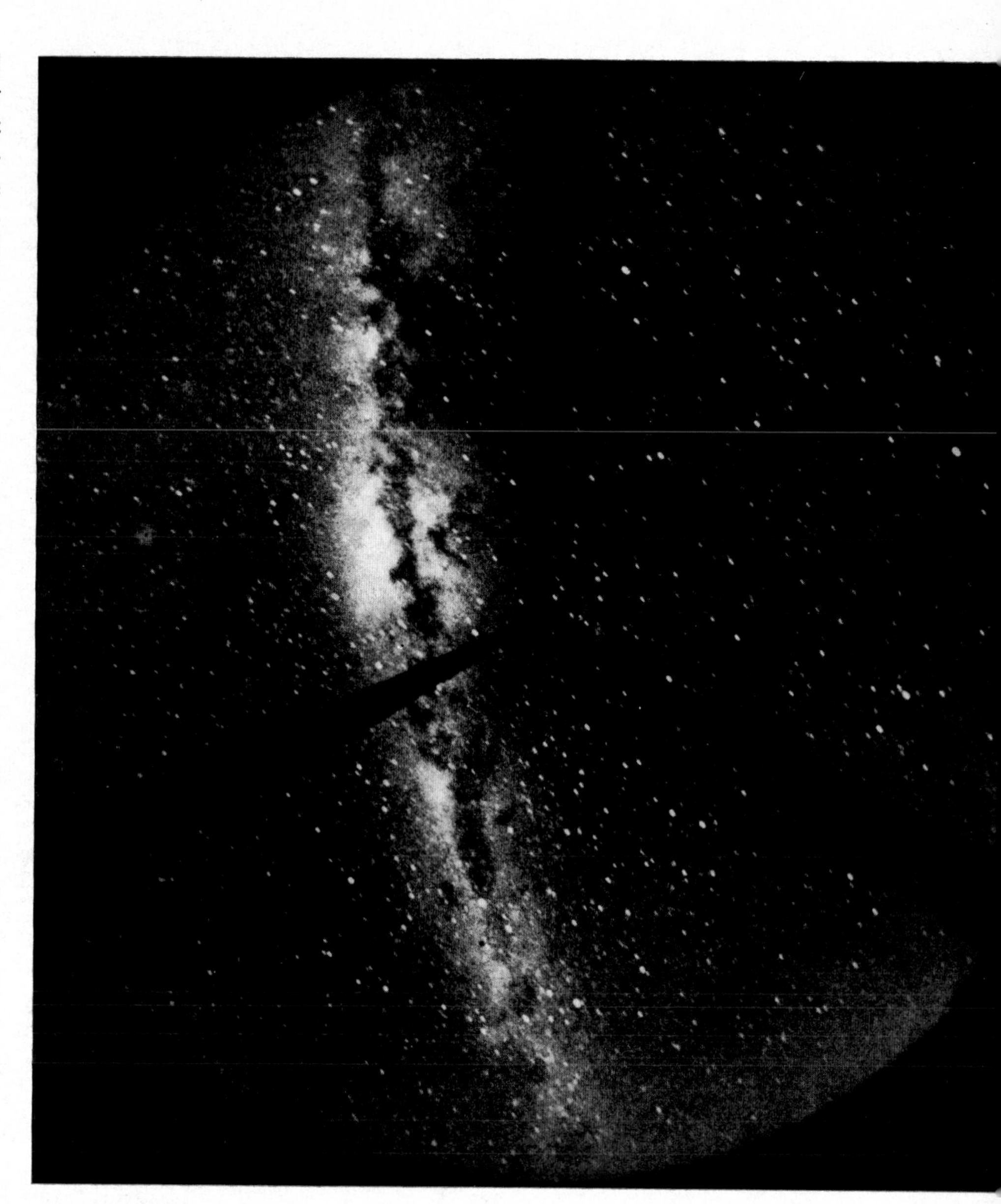

97 On this infrared photograph of the whole of the southern Milky Way, the central region of our galaxy can be seen as a bright cloud in the center of the picture. The clouds of interstellar dust in front of this area form a narrow dark stripe across the bright Milky Way.

From the intensive infrared radiation emanating from the nucleus it appears that a great number of stars is concentrated there, representing together about 30 million solar masses. This would imply that the density of the stars in the inner parts of the nucleus is a million times greater than in the area around the Sun.

It is estimated that the galactic disk has a diameter of 100,000 light-years but its average thickness is only about 1,000 light-years. It is surrounded by the globular star clusters, individual stars and clouds of hydrogen, these occupying a spherical region of 150,000 light-years in diameter.

In accordance with the law of gravitation, the stars and interstellar matter rotate around the center of the Milky Way. The Sun is at a distance of about 30,000 light-years from this and moves at a speed of some 250 kilometers (155 miles) per second on its orbit around the galactic center. Despite this, the Sun needs approximately 230 million years for a complete galactic orbit.

The total mass of our galaxy is roughly equivalent to 130,000 million solar masses, and it is estimated that it comprises about 100,000 million stars. Only two percent of the total mass of the Milky Way system is accounted for by interstellar matter.

In his investigations of the Andromeda nebula in the middle of the 1940's, Walter Baade (1893—1960) discovered that the stars in the spiral arms clearly differ from those of the nucleus region. Whereas the spiral arm structure is largely characterized by stars of high luminosity, the central region mainly contains cooler giants and dwarf stars which also occur in the globular clusters. Baade took account of this fact and divided the stars schematically into two groups which he called "populations." A notable difference between them is expressed, for instance, by their chemical composition. Thus the proportion of chemical elements heavier than helium is about ten times higher in the stars in the spiral arms than in the stars of the central region or the globular clusters.

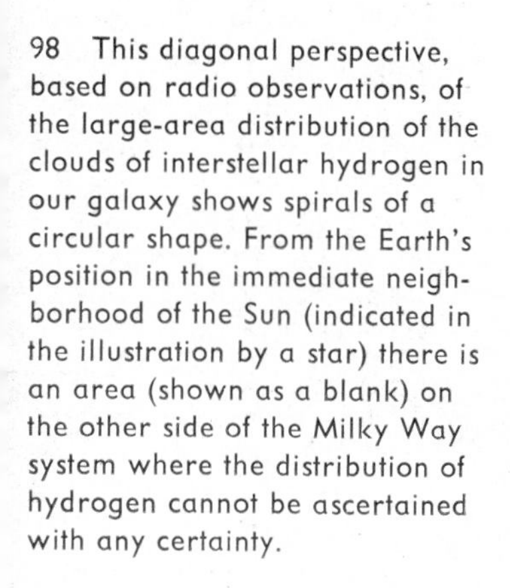

98 This diagonal perspective, based on radio observations, of the large-area distribution of the clouds of interstellar hydrogen in our galaxy shows spirals of a circular shape. From the Earth's position in the immediate neighborhood of the Sun (indicated in the illustration by a star) there is an area (shown as a blank) on the other side of the Milky Way system where the distribution of hydrogen cannot be ascertained with any certainty.

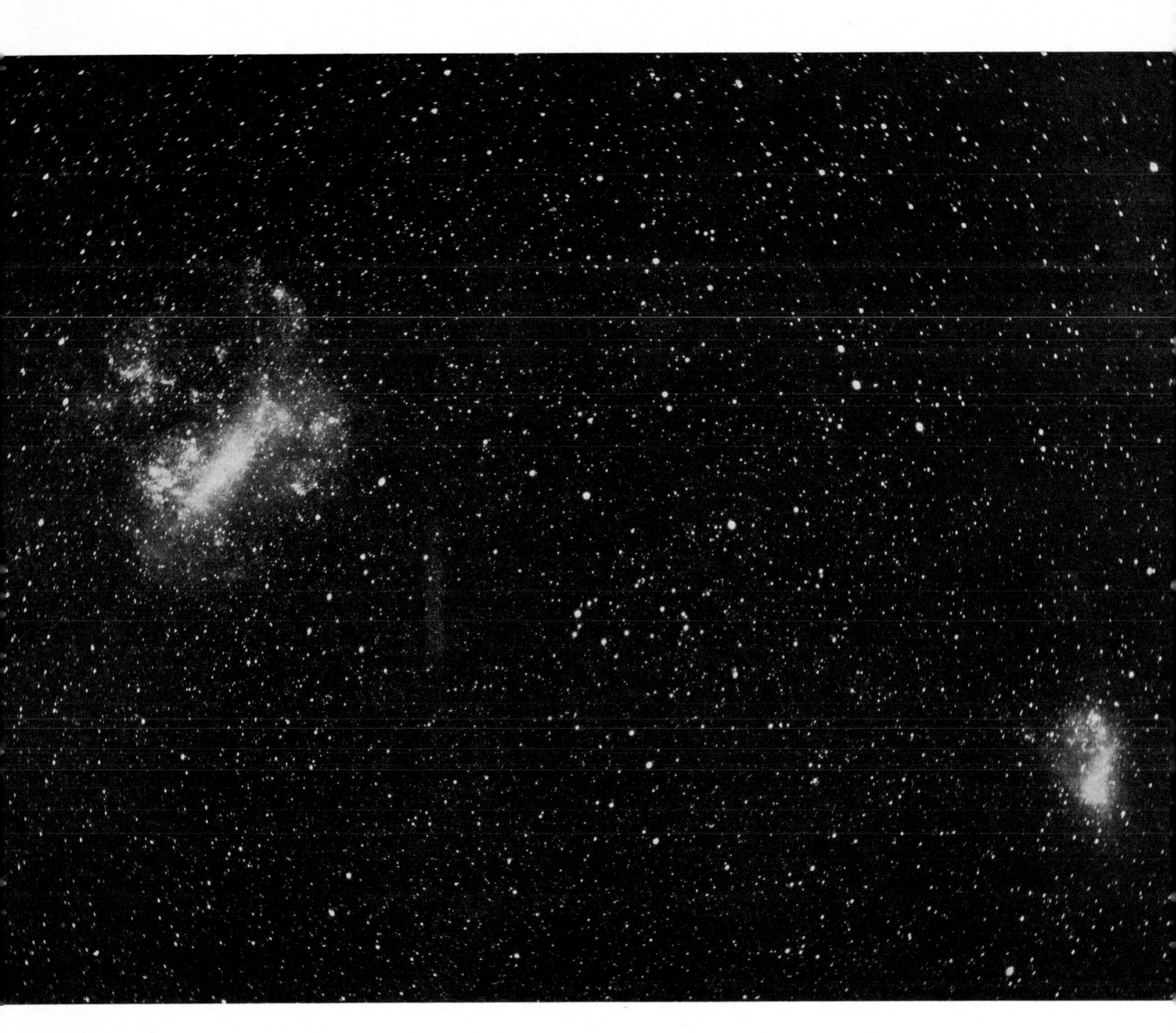

99 The two Magellanic Clouds in the southern sky are small galaxies of irregular shape that form a group of three with our Milky Way system. The Large Magellanic Cloud is about 150,000 light-years away from us, while the Small Magellanic Cloud is somewhat further, approximately 180,000 light-years away. The mass of the former is about 6 million solar masses and that of the latter about 1.5 million.

Like the Andromeda nebula, to which belong at least four dwarf galaxies, our Milky Way system is accompanied by two minor galaxies. These are the two Magellanic Clouds in the southern area of the night sky, named for Ferdinand Magellan, the Portuguese seafarer. Possibly a third companion, newly discovered by means of infrared techniques, the galaxy Maffei I belongs to our galaxy, too.

It is probable that the gravitational effects resulting from these galaxies are responsible for the deviations already indicated in the spatial distribution of the interstellar hydrogen in the outer areas of the Milky Way which otherwise form a flat layer.

Since the Magellanic Clouds are on the doorstep, so to speak, of our Milky Way system, observations of them can be made without much difficulty. Some of the outstanding questions of extragalactic research can even be answered with the aid of medium-size instruments.

Exploring the galaxies

Only by using giant telescopes are astronomers able to advance farther into the world of the galaxies. Long-exposure photographs reveal an astonishing wealth of details and also enable some repetitive structural elements to be detected which form the basis for a simple system of classification evolved by Hubble. Its essential features are still in use and it distinguishes the irregularly formed galaxies from the ones of regular structure, i.e., from the elliptical stellar systems, the normal spiral systems and the barred spirals. Depending on their position in Space, we see them from the front, more or less diagonally, or from the side. Thus there are excellent opportunities in various cases for observing the structural features. In some galaxies, very good observations can be carried out of the distribution of dust in long lanes, following the spirals of the arms. At the same time, it can be seen how the young open star clusters, the luminous gas nebulae and the individual stars of high luminosity are arranged like strings of pearls in the spiral arms.

From time to time, we witness nova or supernova explosions in distant galaxies. Even though these events may only be followed photographically, it is an impressive experience to see a single star flare up within a few hours to a degree of brightness equivalent to the luminosity of an entire stellar system. Tycho Brahe has left us a detailed eyewitness account of the supernova that flared up in the Milky Way system in 1572. He compared the brightness of the new star with that of the planet Venus at the time of the latter's greatest glory. It is even related that it was possible to see the supernova by day and by night, even through fairly thick clouds. The last supernova outburst witnessed by mankind in our Milky Way system occurred in 1604 and it was observed by such celebrities as Galileo and Kepler.

Normally, the radio frequency radiation of the galaxies is much less intense than their visual radiation, but there are also a number of galaxies whose radio radiation is considerably stronger than their visual radiation and these are consequently termed radio galaxies. In these cases, the stellar systems which can be visually observed—and which, incidentally, bear clear traces of gigantic explosions—are flanked by two far-reaching regions from which the radio radiation comes. The spectral characteristics of the radiation emitted by the radio galaxies in the visual and radio range reveal that they owe their origin to electrically charged particles of very high energy which move at high speeds within powerful magnetic fields.

The attempts to identify newly discovered double radio sources of high intensity with optically visible galaxies led, at the beginning of the 1960's, to the sensational discovery of the quasi-stellar objects or "quasars." As indicated by their name, they have a star-like visual appearance but in the radio frequency range they behave like the radio galaxies. The spectra taken from quasars were a special kind of riddle for astronomers. At first, the lines occurring in them were incompatible with any interpretation since no chemical element could be associated with these spectral lines. It was only after some unsuccessful attempts to explain them that it was noticed that all the lines have a large red shift. Meanwhile optical observation methods have been used for searching quasars. With these methods a lot of new findings were made. It was surprising to find out that a very high percentage of the newly discovered quasars is radio-quiet, i.e., does not emit radio waves.

We are still a long way from a final explanation of the new problems which have arisen with the discovery of the quasars. In the meantime, it is only known that the quasi-stellar radio sources are bodies which have to be described with superlatives. It is likely that the quasars, in a relatively small space, represent a mass equivalent to that of entire stellar systems. The small, high-mass nucleus is enclosed in

100a—c The star system about 14 million light-years away in the constellation Canes Venatici possesses, from our point of view, a very fine spatial position, enabling us to follow exactly the course of the two clearly defined spiral arms (a). One of them forms a bridge with the companion galaxy. The long dark lanes on the inner edges of the spiral arms are produced by interstellar dust. As can be seen from the drawing (b) published by Sir John Herschel in 1833, the spiral structure of the "nebula" had escaped him. This was first noticed by the Earl of Rosse, as apparent from the drawing (c) which he published in 1880.

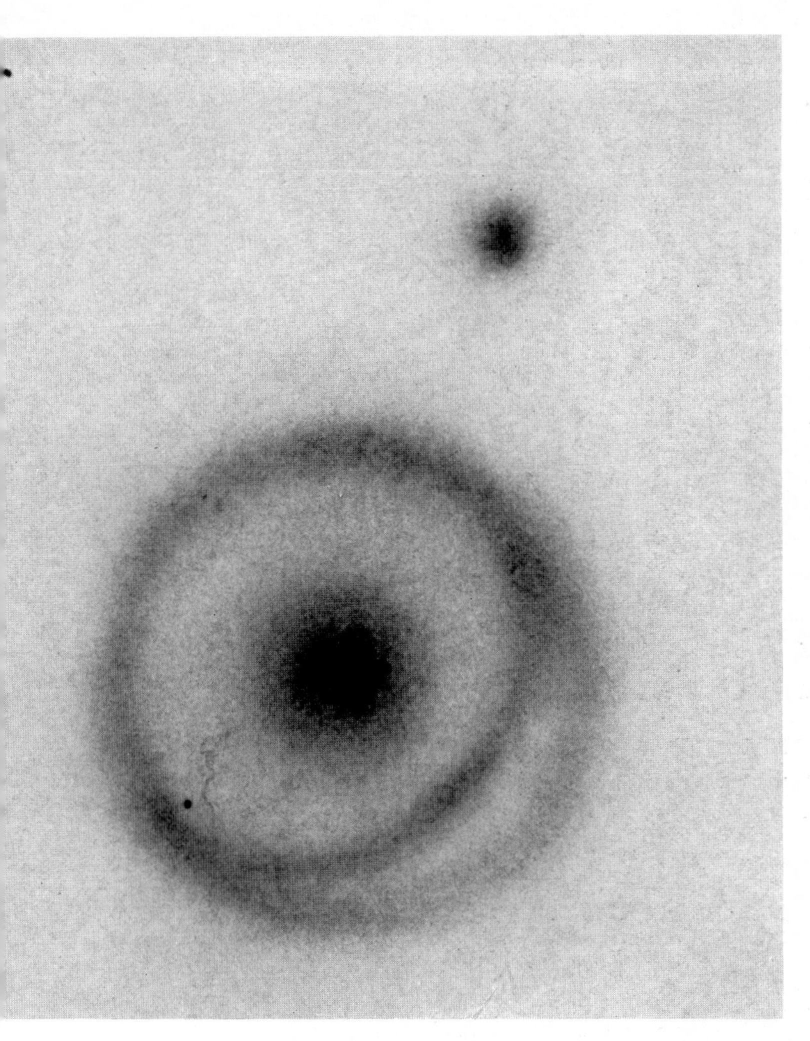

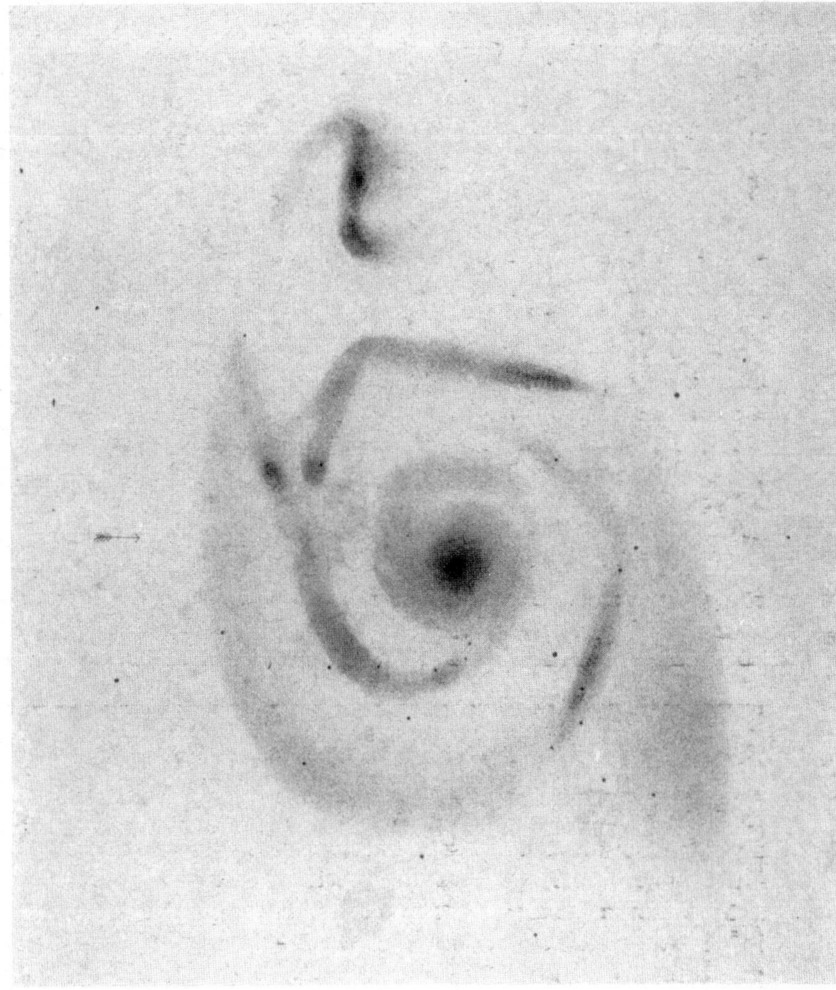

a gas envelope which, in terrestrial terms, is extremely thin. On the outside of this there are the two extensive regions which emit the radio radiation. Its specific characteristics indicate that every quasar has suffered a mighty explosion within the cosmically recent past, i.e., a few hundred thousand years ago or at most within a hundred million years. It is suspected that a promising relationship exists between the quasars, the exploding galaxies and the radio galaxies.

The development of the Cosmos

If another look is taken at the past, there is yet another discovery by Hubble which claims an outstanding place in the investigations of the 1920's. This is the proof of the fact that the general red shift of the spectra of the galaxies (first discovered by the American V. M. Slipher) increases linearly with distance. There is no longer any doubt that the red shift is to be interpreted as the outward movement of the galaxies in an expanding Cosmos. At first sight, our Milky Way system appears to occupy a special position in it similar to that of the Sun in Copernican astronomy. It should be realized, however, that in an expanding Space every stellar system must move away from every other. Consequently, the same impression is obtained at any point of observation in Space, i.e., that one is at the center of the process of expansion.

Shortly after the quasars were found, another important astronomical discovery was made. From measurements of radio waves, a weak radiation was

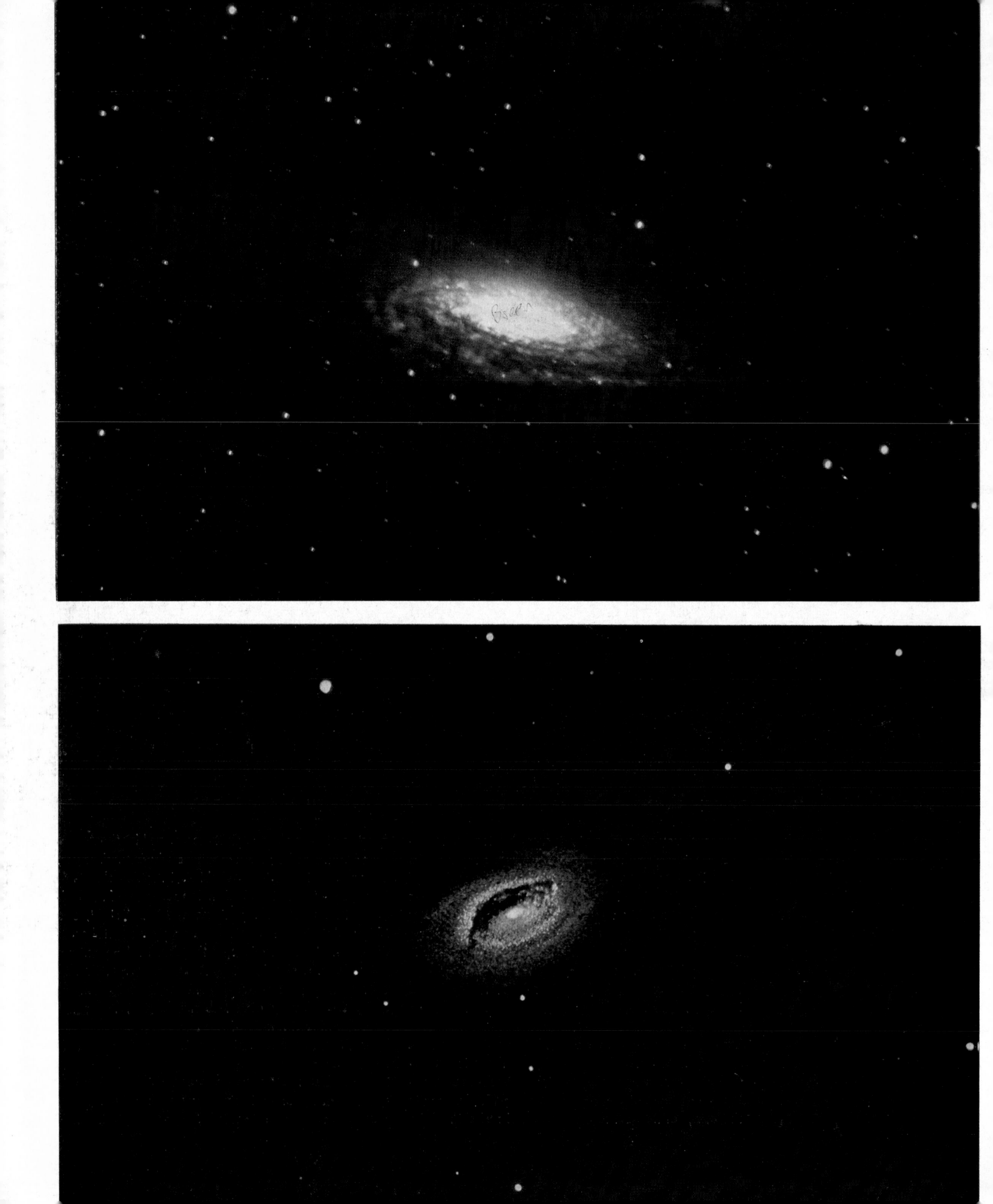

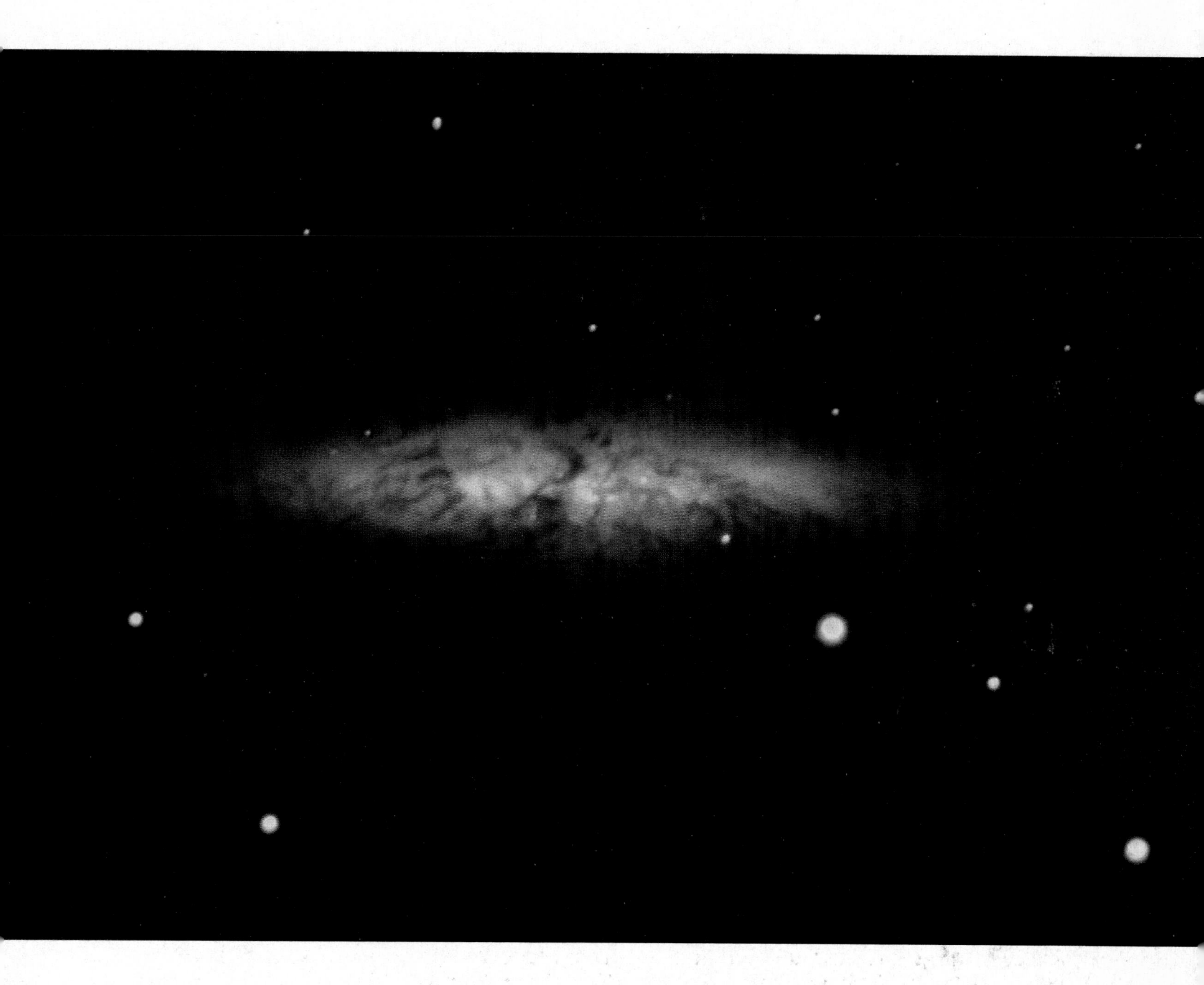

101 a, b The two Baade star populations (that is, all stars which in respect of age, chemical composition, spatial arrangement in the galaxies, and their motions are similar to each other) can be clearly distinguished on the color pictures of the spiral star systems NGC 7331 in the constellation Pegasus (a) and NGC 4826 in Comae Berenicis (b). While the yellowish-red coloring in the central regions of the galaxies is caused by the older stars of Population II, the spiral arms are somewhat blue in color, due to the very hot young stars of Population I.

102 Between one and three million years ago, a colossal explosion occurred in the heart of the M 82 stellar system. The consequence of this was that clouds of hydrogen were expelled from it at speeds of about 1,000 km/s. The total mass of the red hydrogen clouds is of the order of 6 million solar masses. Furthermore, powerful radio radiation is emitted by the galaxy, this being caused by the interaction of rapidly moving electrons and a strong magnetic field. The young hot stars account for the blue coloring of the outer parts of the stellar system.

103a—e The comparison of different types of galaxies gives a good idea of the differences in their structure. Some galaxies have practically no structure, while others are of a marked spiral form. In the barred spirals, the spiral arms start to coil only at a point far outside the nucleus of the galaxy. We see the spiral

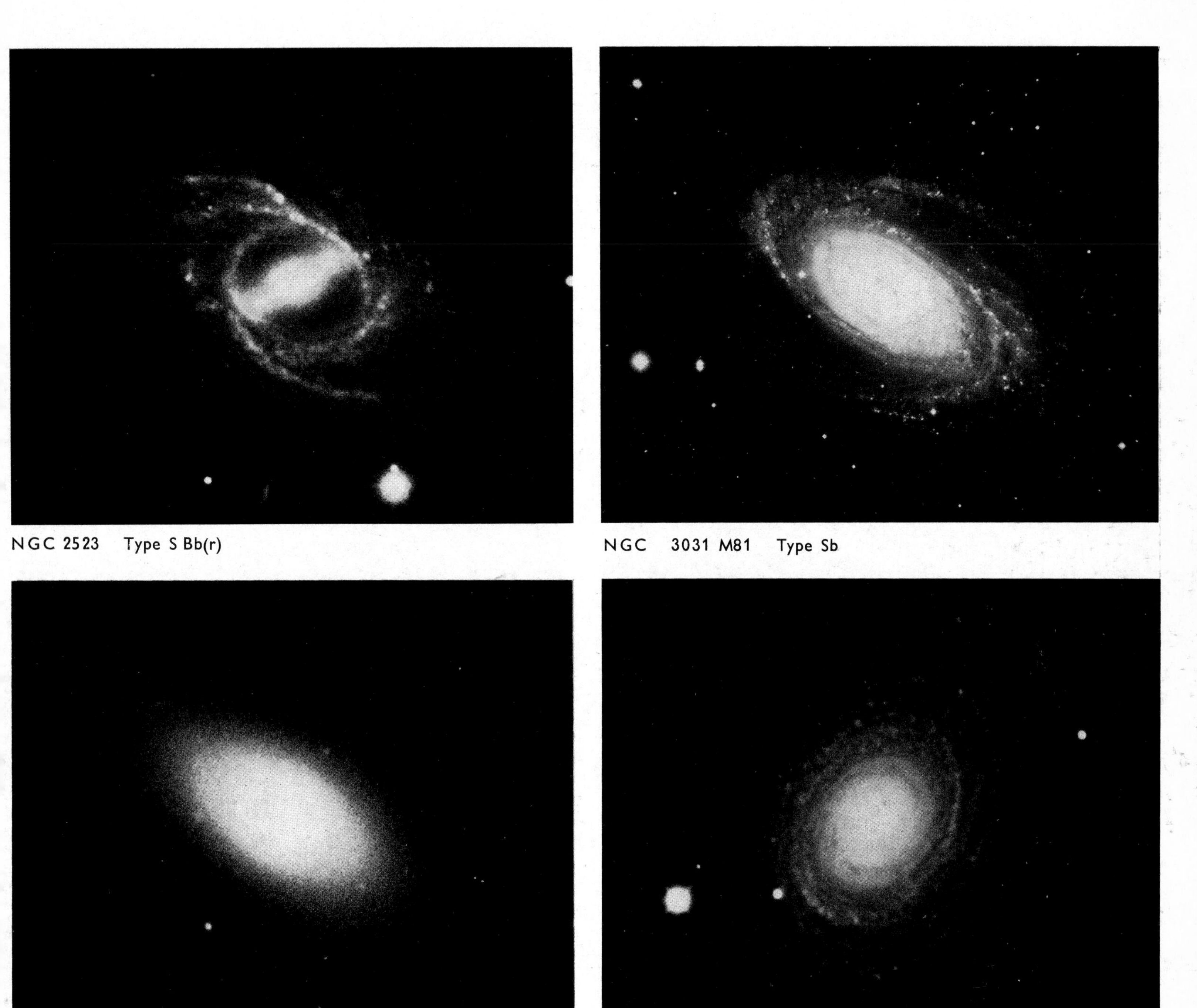

NGC 2523 Type S Bb(r)

NGC 3031 M81 Type Sb

NGC 1201 Type SO

NGC 488 Type Sab

stellar system M 104 in the constellation Virgo almost exactly from the side (left page). The dust concentrated in the galactic plane of this stellar system screens off the light from the stars and appears to divide this galaxy into two parts.

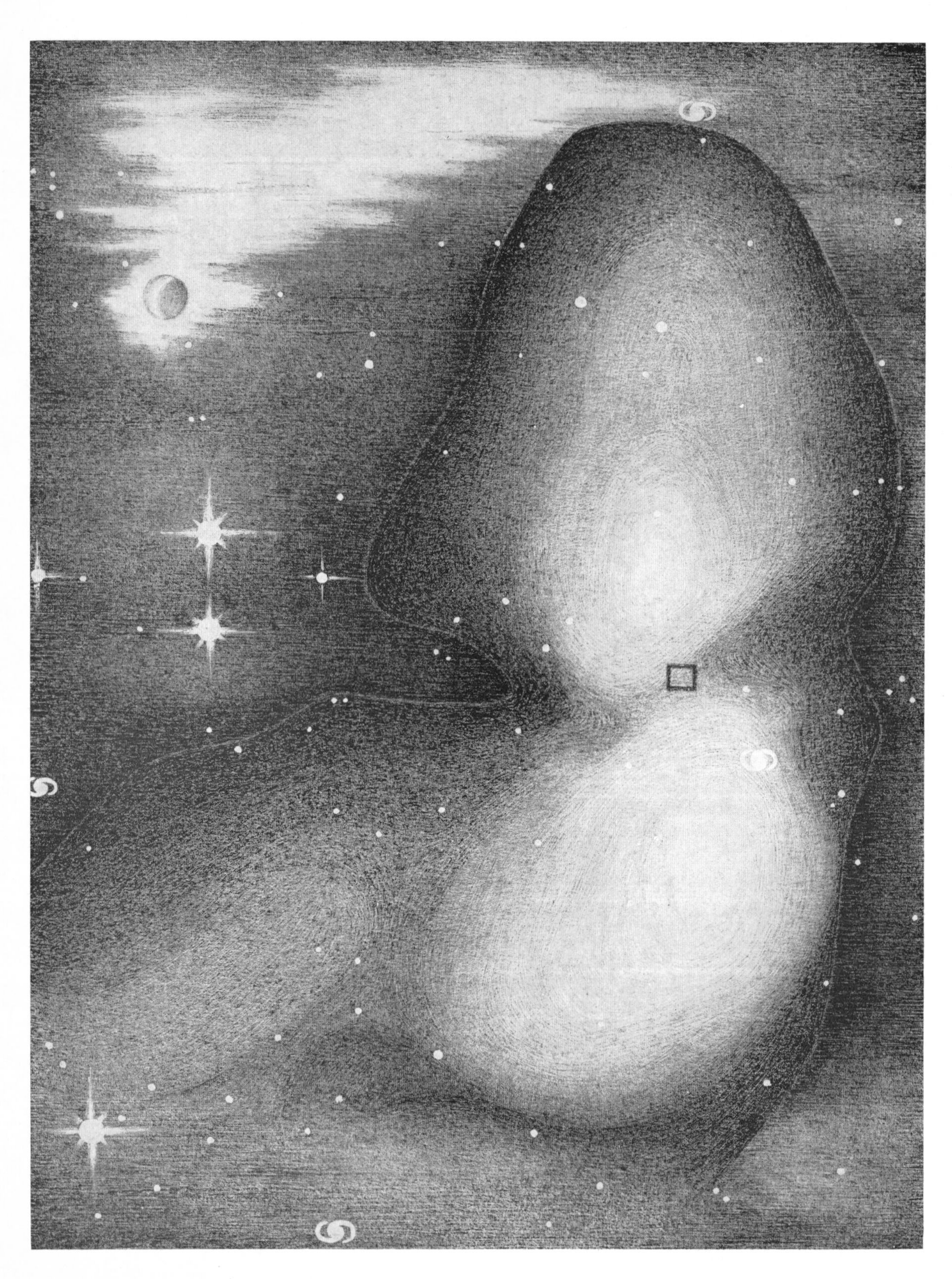

104a, b When the intensity of the radio radiation is converted into magnitudes, a remarkable picture is then obtained of the radio sky in the surroundings of the NGC 5128 spherical stellar system in the constellation Centaurus. The radio radiation comes from two very large regions, between which there is the stellar system which can be observed visually and is some 14 million light-years away. In the drawing (a), the rectangle gives an accurate idea of the scale of the photograph (b), and the apparent diameter of the Moon provides an impression of the large angle of the radio source. The dark belt of interstellar dust stands out against the visible stellar system. A violent explosion must have taken place here a few million years ago.

identified which came from every direction. Its spectral characteristics indicate that it is comparable with the black body radiation known in physics. In this sense, the Cosmos emits radiation which would be equivalent to that of a black body with a wall temperature of three degrees above absolute zero.

It is worth noting that both the general outward movement of the galaxies and the 3 Kelvin radiation had already been predicted on the basis of theoretical considerations. Both effects demonstrate that Space as a whole has a history.

However, the development of the Cosmos in the past can only be traced on the theoretical level. Taking into account all the uncertainties which result from extrapolations of this kind, it appears that the Cosmos possessed quite different characteristics about 15,000 million years ago. It was much smaller than it is today, and for a short time temperatures of more than 10,000 million degrees were found within it. Under these conditions, the primordial material of the Universe was in an unusual state. This material, i.e., elementary particles with rest mass, and the radiation consisting of photons, that is, elementary particles without rest mass, was in a state of equilibrium, forming a single mixture. In this state, radiation was the principal component in the energy balance of Space. The expansion which took place at an explosive rate led to a rapid dilution and cooling down of this mixture, thus disturbing the balance between matter and radiation. The two decoupled and went their own ways at temperatures of not more than a few thousand million degrees. In contrast to matter, radiation has experienced and is experiencing not only dilution, since Space is constantly expanding, but also a reduction in energy since its wavelength is constantly increasing in the course of expansion. Through this "inflation," it is becoming less and less significant while matter is becoming the dominating component of the universe. In the course of the transition from cosmic Space filled with radiation to the Universe existing at the present time, which is composed mainly of elementary particles with rest mass, the first cosmic frequency distribution of chemical elements also appeared. Observations show that the expansion of the Cosmos is still continuing even now. The remainder of the radiation from the hot phase of the Cosmos is found in the form of the 3 Kelvin radiation.

The stellar systems were probably formed more than ten thousand million years ago, but how this process took place is still not clear. Basically, it is a question of two possible ways. It is conceivable that extensive, high-mass clouds of gas were drawn together by the action of their own gravity and were compressed to spherical forms. In the course of contraction, the clouds broke up into smaller parts in which contraction continued to take place. This was again interrupted by a process of disintegration which in turn was followed by another contraction phase. When the clouds of gas had been divided in this manner into many small and fairly dense individual parts, the way was probably clear for the actual formation of the stars. The globular clusters may have been the first to be formed in this way, and the stars populating the central regions of the stellar systems may also have come into existence at this time. It was only later that the formation of the younger stars took place in the outer areas of the galaxies.

In the hypothesis just outlined an initial rotational movement of the clouds of gas plays a great role. Since their angular momentum is not lost, a flattening of the stellar systems also takes place during the contraction, as may be observed in many cases.

According to the other possibility, the formation of the galaxies was initiated by a "discharge" of mass from a super-dense body of small size, this being accompanied by events of an explosive nature. The expelled clouds subsequently changed into stars. This hypothesis considers a stellar system as developing from the inside outward, as it were.

Unfortunately, the theoretical concepts of the emergence of stellar systems do not yet permit any conclusions to be drawn regarding the appearance of their special forms. Until now, the only promising calculations carried out have been in connection with the formation of the spiral arms. According to these calculations, the effect of gravitation necessarily led to a spiral pattern being formed from the initially totally unordered distribution of the stars in the galactic disk, this pattern then being maintained. The formation of the elliptical galaxies, however, has not yet been explained. It is only certain that they are neither a very early nor a very late phase of development of the spiral stellar systems.

The emergence and development of stars

In the description of the various stages of development of the galaxies, as this is now understood, only brief mention was made of the emergence and development of the stars. Let us now go into further detail.

In the course of the process of disintegration of the great cloud complexes already mentioned, clouds were formed of smaller and smaller size and less and less mass from which, at the end of a further process of contraction, something qualitatively new ultimately emerged, namely, stars. From certain observations, it can be concluded that as the large clouds disintegrate, very many more star-forming clouds of low mass result than of high mass. This explains the fact noted earlier in connection with the physical parameters of the stars that a relatively low number of high-mass, hot stars contrasts with a great many low-mass and fairly cool stars.

So far, the formation of stars has been the subject of discussion. The following remarks will concentrate on the physical processes in the interior of the stars during the various stages of development. The basic form of every star may be regarded as a wide-ranging and practically transparent cloud of gas mixed with particles of dust, shrinking under the action of its own gravity. In the course of time, a dense center is formed which constantly grows at the expense of the outer regions of the star-forming cloud. Because of the constant increase in density of the contracting cloud, radiation can no longer freely escape from its center; the thermal energy resulting from the compression of the cloud can no longer be emitted without hindrance. Only about half of it reaches the outside while the other half serves to raise the temperature of the emerging star, the protostar. This process recalls the views of Helmholtz, mentioned at the beginning, who tried to explain stellar development as a whole as a continuing process of contraction. Modern research findings have shown, however, that the contraction of the star-forming clouds is only a relatively short intermediate stage in the entire cycle of stellar evolution.

When the contraction of the protostar is so far advanced that the temperature within it has risen to several million degrees, a sharp increase takes place in the number of energy-supplying nuclear reactions. The energy released serves largely to raise the interior temperature of the star still further and, like a bellows, fans the fire within the stellar nuclear reactor. As a result, the temperature and also the pressure within the star become higher. This feedback process of increased temperature and pressure gradually halts the contraction as a result of the continual rise in the amount of energy released. At the same time, temperature and pressure also stop rising. In this way, the star begins the longest phase of its life, during which there is scarcely any change in its internal composition and physical parameters.

105 In the cluster of galaxies in the constellation Hercules, the most diverse types of galaxies may be seen in close proximity to each other. The smaller of them are distinguished from the stars in our Milky Way system only by their faded appearance. There are between several hundred and several thousand stellar systems in the clusters of galaxies. The one in the constellation Hercules is about 300 million light-years away.

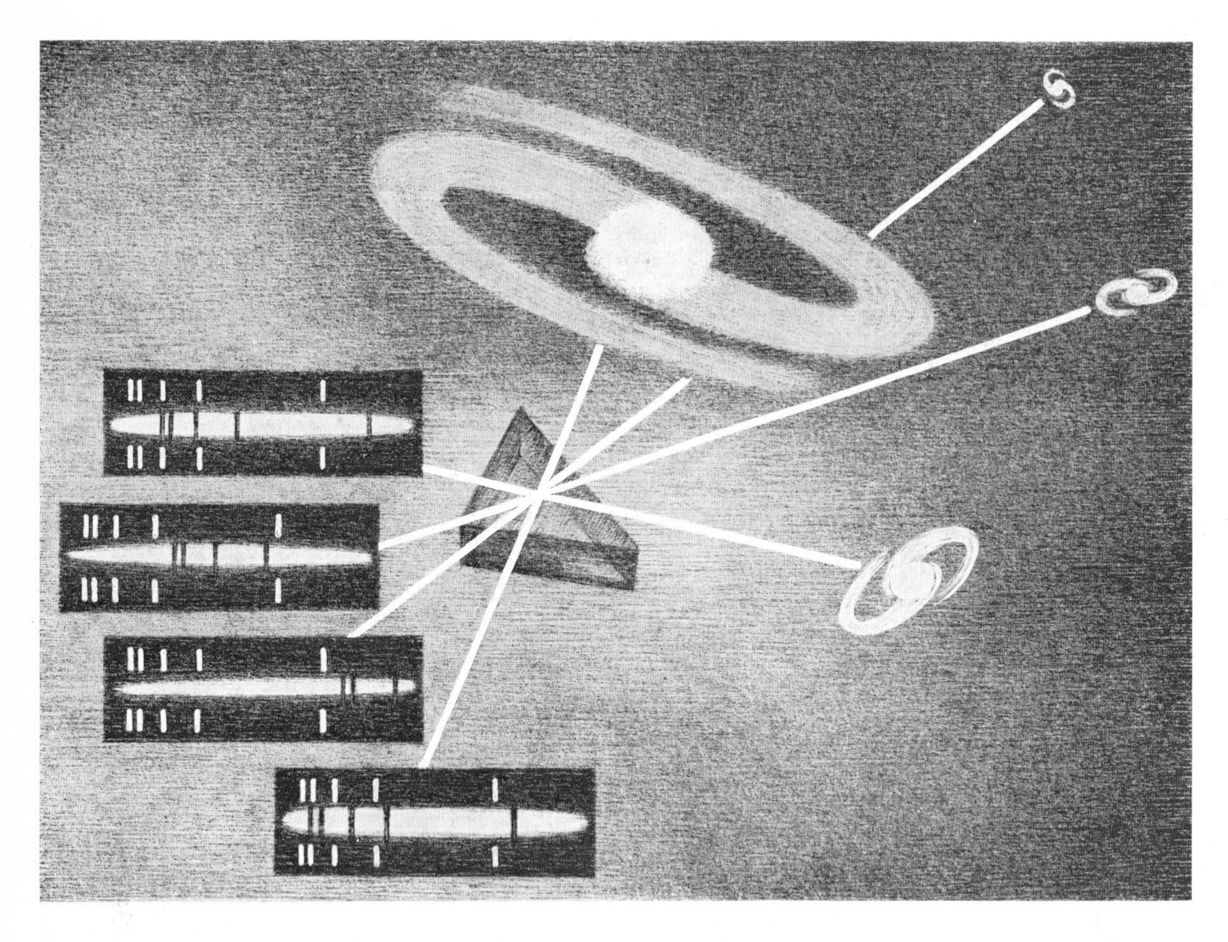

106 The outward movement of the galaxies is reflected in the red shift in their spectra. This can be measured by comparing lines that have not shifted with the absorption lines in the spectra of the galaxies. Discovered by Hubble, this effect states that the red shift of the spectra increases linearly with the distance of the stellar systems. The greatest line shifts measured correspond to outward velocities of about 150,000 kilometers per second, i.e., about half the speed of light.

Through the energy-supplying nuclear transformation of hydrogen into helium, the central zones are enriched with helium in the same proportion as the sources of raw material decline. When, in the course of time, the reserve of hydrogen in the original combustion zone is exhausted, the "hydrogen burning" shifts to a spherical shell around the continually growing helium nucleus. The frequency distribution of chemical elements emerging in the stellar interior as a consequence of the nuclear reactions necessarily leads to changes in the overall structure of the star. These take place imperceptibly but continuously over a long period of time.

How long a star remains in this stage of development, which is characterized by changes scarcely perceptible from the outside, depends solely on its mass. Thus we return to the relationship already briefly described between the mass and the luminosity of stars which is responsible for the apparently paradoxical fact that the high-mass stars have been the quickest to convert their reserves of hydrogen to helium. Consequently, stars with surface temperatures of about 35,000 Kelvin can only remain in the stable state described for seven to ten million years. On the other hand, for example, the Sun with its lower mass will only have exhausted its hydrogen-based reserves of energy after more than 10,000 million years.

When the hydrogen reserves in the interior of a high-mass star are used up and the resulting helium nucleus accounts for about ten percent of the mass of the star, the star undergoes changes of a really far-reaching nature. A relatively rapid contraction takes place in the helium nucleus while its outer layers expand at the same time. In this way, the star changes into a giant star of lower surface temperature but of high luminosity. Its principal sources of energy are still the hydrogen processes, now taking place in a spherical shell. If, however, in the course of the contraction of the helium nucleus, the temperature

107 How stellar systems are formed has not yet been explained. Perhaps they are produced by the contraction of large clouds of gas. (See the illustrations in the left-hand column.) If the clouds possessed an initial rotary motion, they would flatten out on contraction into disks as a consequence of the spin they still have. The first stars were probably formed in the globular star clusters. Another possibility, represented by the illustrations in the right-hand column, is that stellar systems were formed by an explosive ejection of mass from a super-dense body, perhaps from a quasar. This would imply that exploding galaxies are early stages in the formation of galaxies. The process of the actual formation of stars is essentially the same as in the first hypothesis.

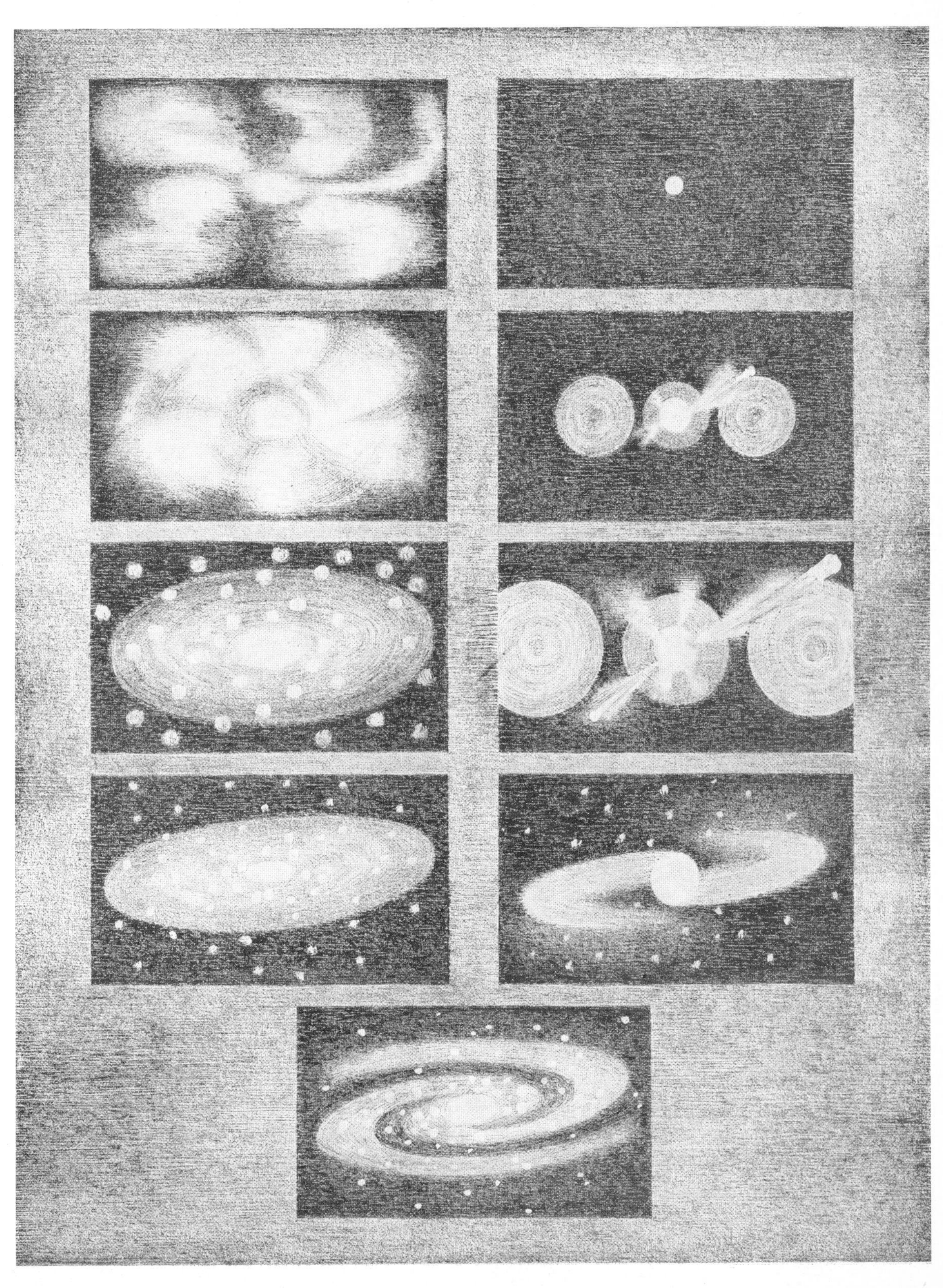

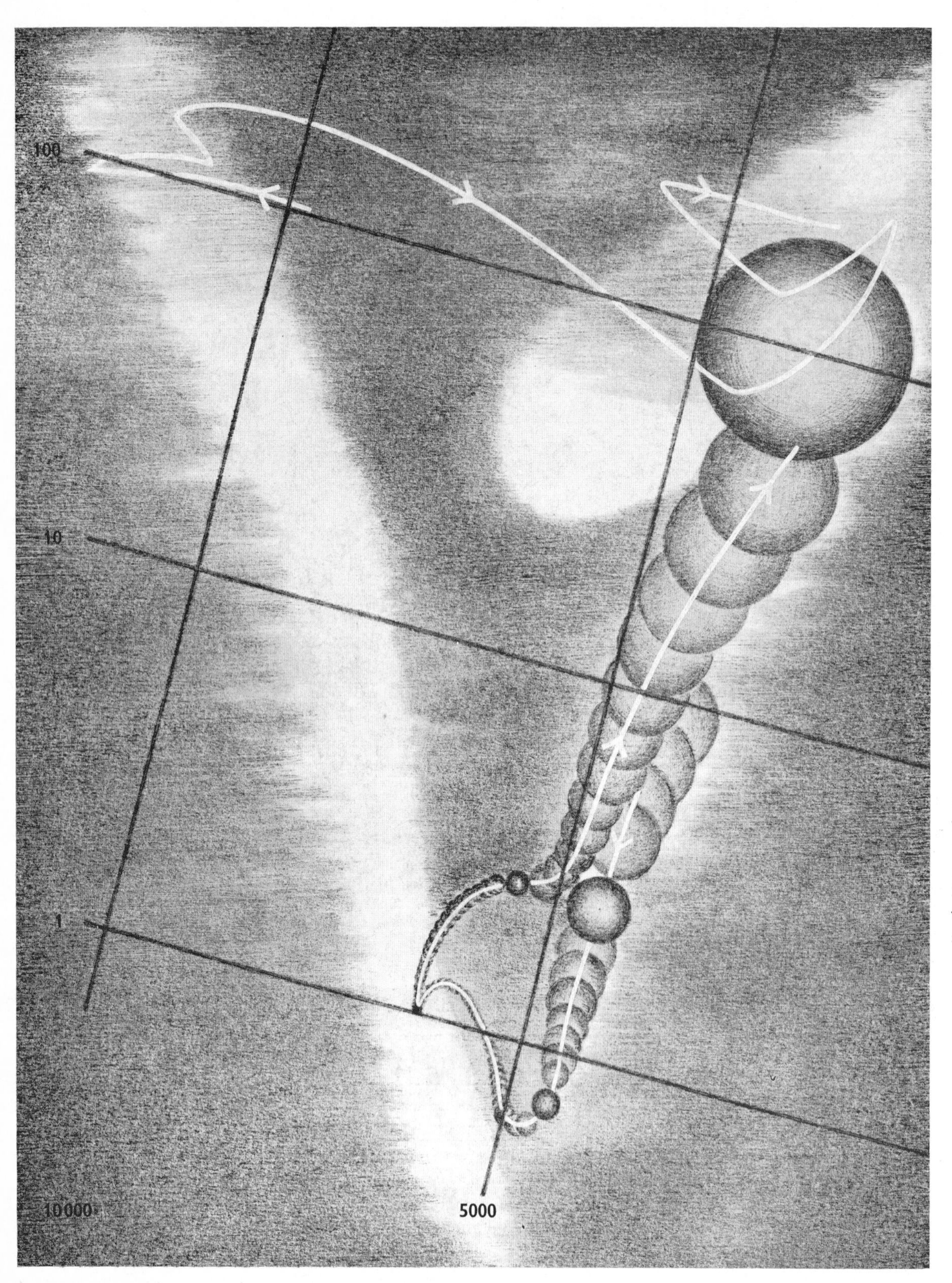

108 In the course of the development through which the stars and the Sun pass, a characteristic change takes place in their internal structure and in their physical parameters. If, for instance, the physical characteristics which the Sun has possessed in the course of its development, possesses at the present time or will receive in the future are entered in the Hertzsprung-Russell diagram, an idea is obtained of its past and future history. Some 5 thousand million years ago, our Sun was still a comparatively large star of reddish luminosity (stage 1, see lower part of illustration, follow the direction of the arrow). In the course of a few million years, this primordial Sun contracted (stage 2), and, finally, about 5 thousand million years ago, achieved the physical state (for instance the size and the surface temperature) it still has today (stage 3). In another 5 thousand million years, during its "ageing process," taking a few million years, the Sun will develop into a giant star of reddish luminosity (stages 4, 5). In the upper part of the illustration, a schematic outline is given of the course of development of a star with five times the mass of the Sun. The luminosity of the Sun is taken as a unit for its luminosity. For the five stages specially marked in the course of development of the Sun, details of the physical relations in the interior of the Sun are shown in the following illustration.

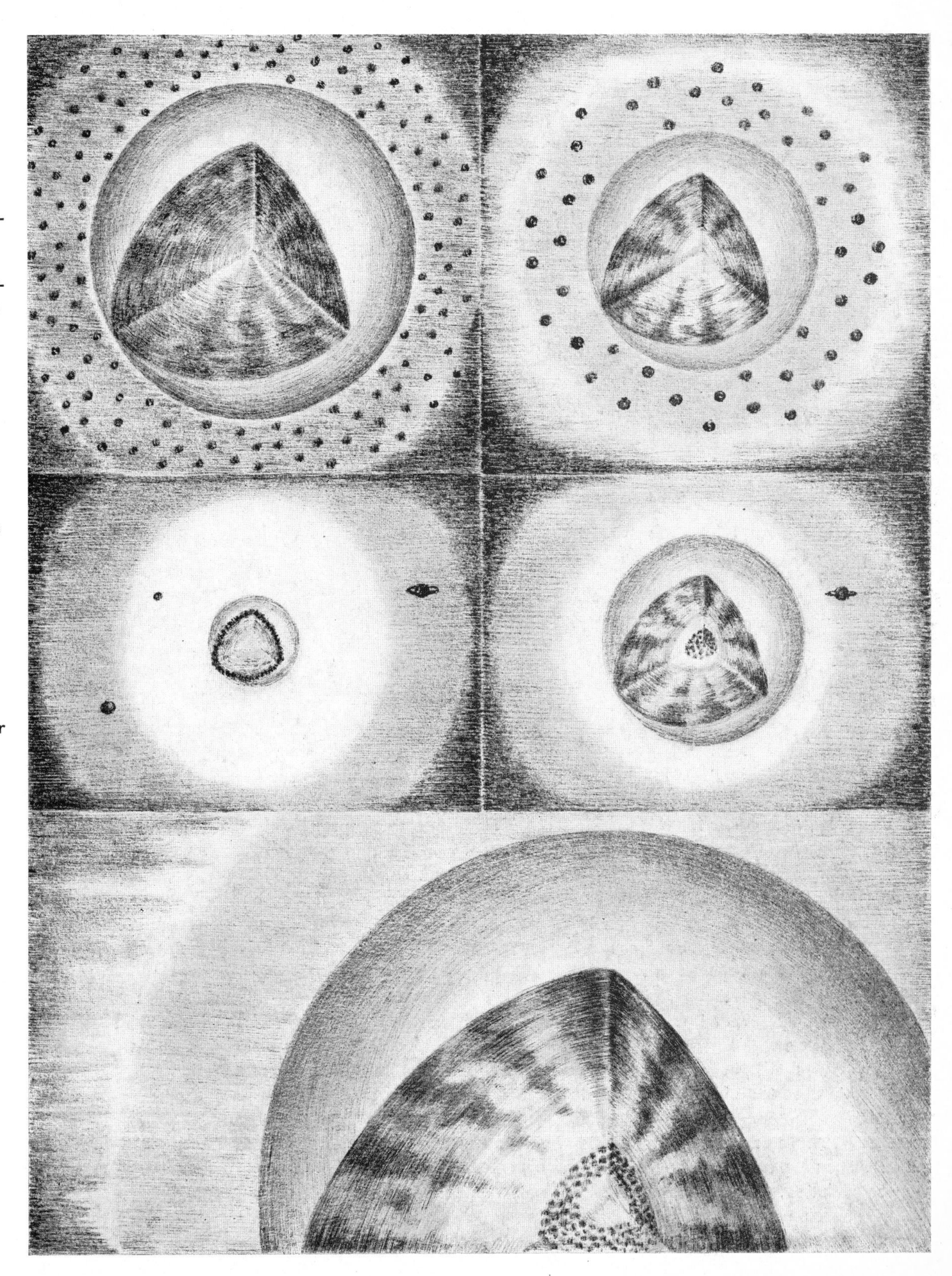

109 The five phases taken from the life-story of the Sun give an idea of its interior in the past, present and future. For reasons of space, the relative diameters have had to be reduced. In the depiction of their future evolution, the central parts of the Sun are exaggerated in relation to their proportional diameter in each phase. About 5 thousand million years ago, our Sun was a large contracting star, interspersed with streams of gas (this interspersion being a mechanism that is called convection). The space surrounding it was filled with gas and many tiny particles of dust (stage 1). In the course of further evolution, the zone affected by streams of gas became smaller and moved away from the center (stage 2). At the present time (stage 3), it comprises only a relatively thin shell under the surface. In the course of thousands of millions of years, the solid bodies of the solar system were formed from the particles of dust. After a few thousand million years, when the store of hydrogen in the center of the Sun is exhausted, it will become a giant star in which the burning of helium will be a new source of energy (stages 4 and 5).

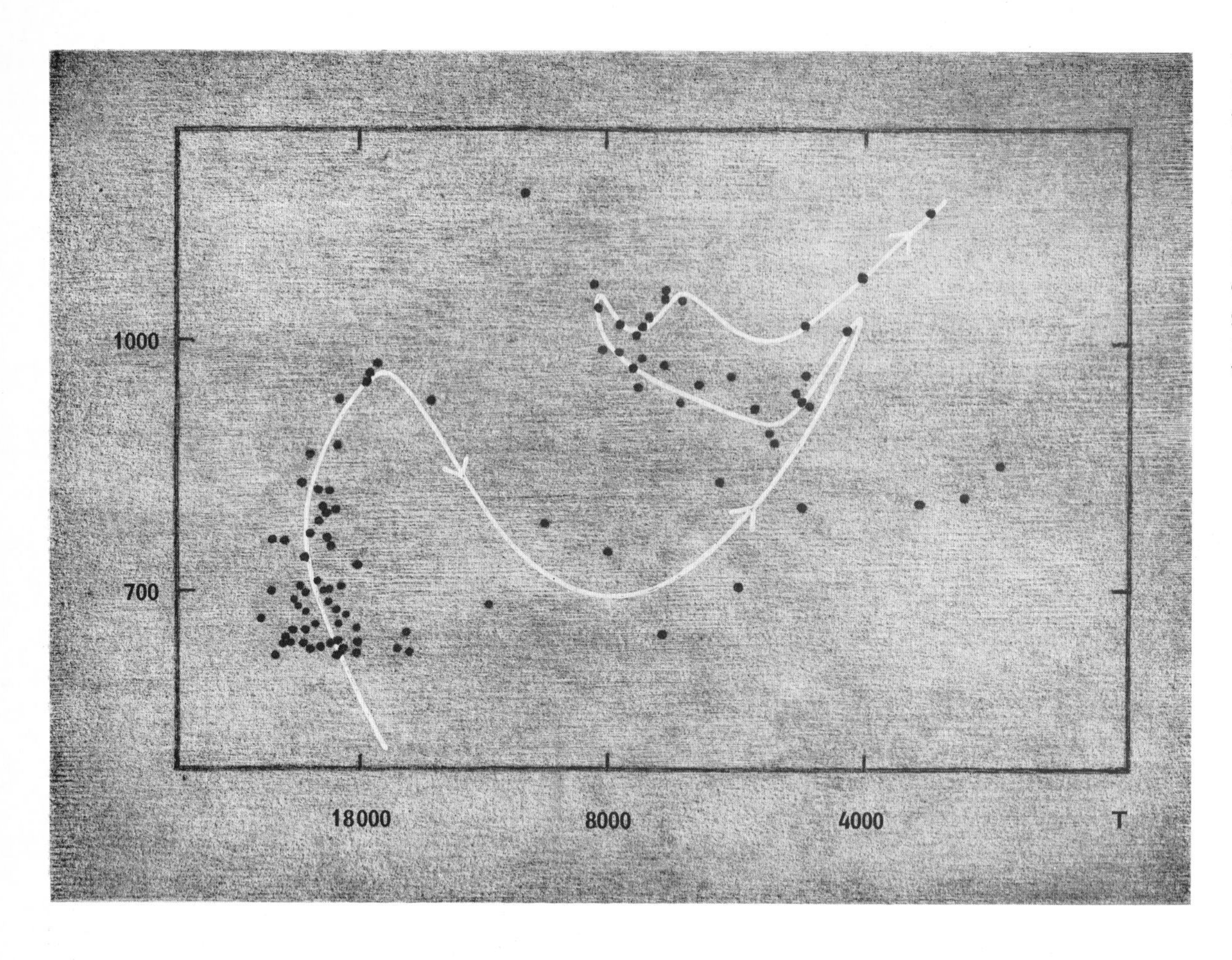

110 In this excerpt from the Hertzsprung-Russell diagram of the open star cluster NGC 1866, all the high-mass stars in an advanced stage of development are entered as points. The continuous line marks the theoretically calculated course of evolution of a star of five solar masses. There is quite good agreement between the theoretical conclusions and the observations made.

exceeds a limit of about 80 million Kelvin, new nuclear reactions occur and lead to the formation of carbon nuclei, the energy for this coming from the burning of the helium nuclei. When eventually the helium reserves are at last exhausted, the helium burning likewise shifts to a spherical shell around the carbon nucleus. It is feasible that it begins to contract and that through yet another rise in temperature a new source of energy is tapped, this being carbon burning. Nevertheless, at the present time we still have only a very incomplete idea of these stages of stellar evolution.

The last-mentioned phases of development take place within a period of not more than ten million years and are accompanied by permanent changes in stellar structure and physical parameters. Due to the enormous expansion of the star in its advanced state of evolution, even a slight overheating of the stellar atmosphere can cause a flow of gas into the Space around it. When the resultant loss of mass is high enough, a very small dense star, a white dwarf, can form from the remaining mass.

In the late stages of development of the stars with a fairly high mass, a nova or supernova outburst is also possible. When this happens, a considerable part of the mass is hurled into Space as if by an explosion. It seems likely that a white dwarf or a neutron star forms from what is left over of the star. These small and very dense objects represent the final state of stars.

From the stellar matter cast into interstellar Space, new cloud complexes are formed again and, under certain circumstances, can lead to a new generation of stars. The strikingly close spatial proximity of interstellar matter in the form of gas and dust and young stars, as, for instance, in the Orion nebula, support the contention that stars were formed in these regions within the cosmically recent past.

If the changes in the measurable quantities taking place in the course of evolution are entered in a

111 A process of importance for the release of atomic energy in the high-mass stars is the carbon-nitrogen cycle which is shown here in somewhat simplified form. In the course of this (starting in the upper left corner of the illustration), one carbon nucleus (the taller of the sketched men) picks up four hydrogen nuclei (the smaller of the sketched men) in succession which leads, in addition to the formation of instable nuclei, to the emission of energy. At the end of the cycle, which will be run through in counter-clockwise direction in the illustration, a helium nucleus becomes detached from the nitrogen nucleus which has formed, and moves away. The carbon nucleus which results from this can then repeat the cycle of nuclear reactions outlined.

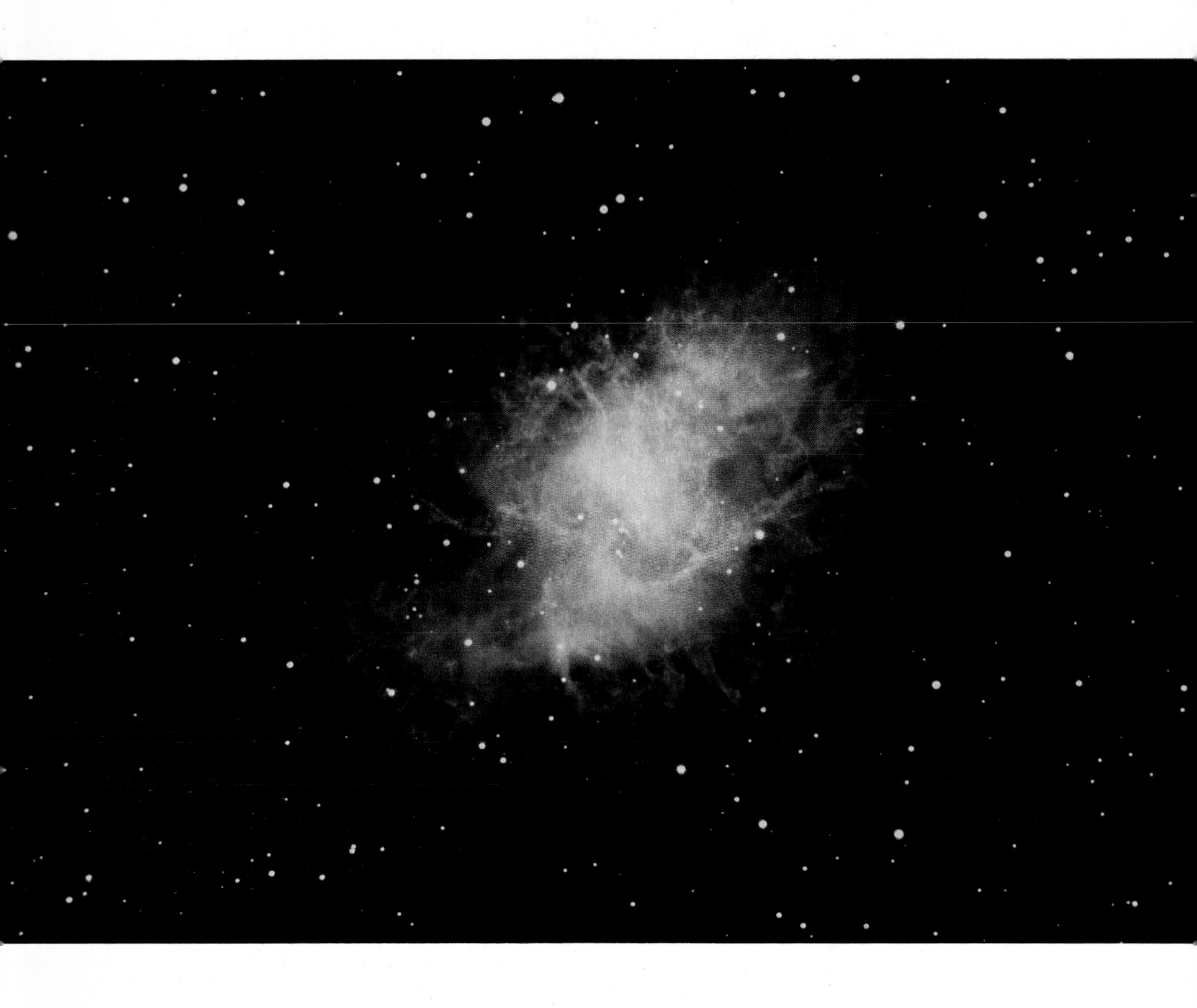

112a—c The Crab nebula in Taurus owes its existence to a supernova explosion which, according to old Chinese records, occurred in 1054. The envelope of the nebula, which is pierced by red luminescent clouds of hydrogen, is still continuing to expand. The white light is caused by high-speed electrons moving in a powerful magnetic field (color photograph). The Crab nebula emits not only intensive radio radiation but even X-rays. Interestingly enough, there is a pulsar in it. The question as to which of the stars was the one that actually exploded has only recently been answered. A new observational technique was successfully used to obtain visual proof for the first time of the radiation flashes coming from the pulsar. The illustration (b) shows this star in its bright (top) and dark (bottom) phases. Since pulsars probably owe their origin to a supernova explosion, this identifies the star in question. The crab form of the nebula is clearly shown in Littrow's *Atlas des gestirnten Himmels* (c, next page) which was published in 1867.

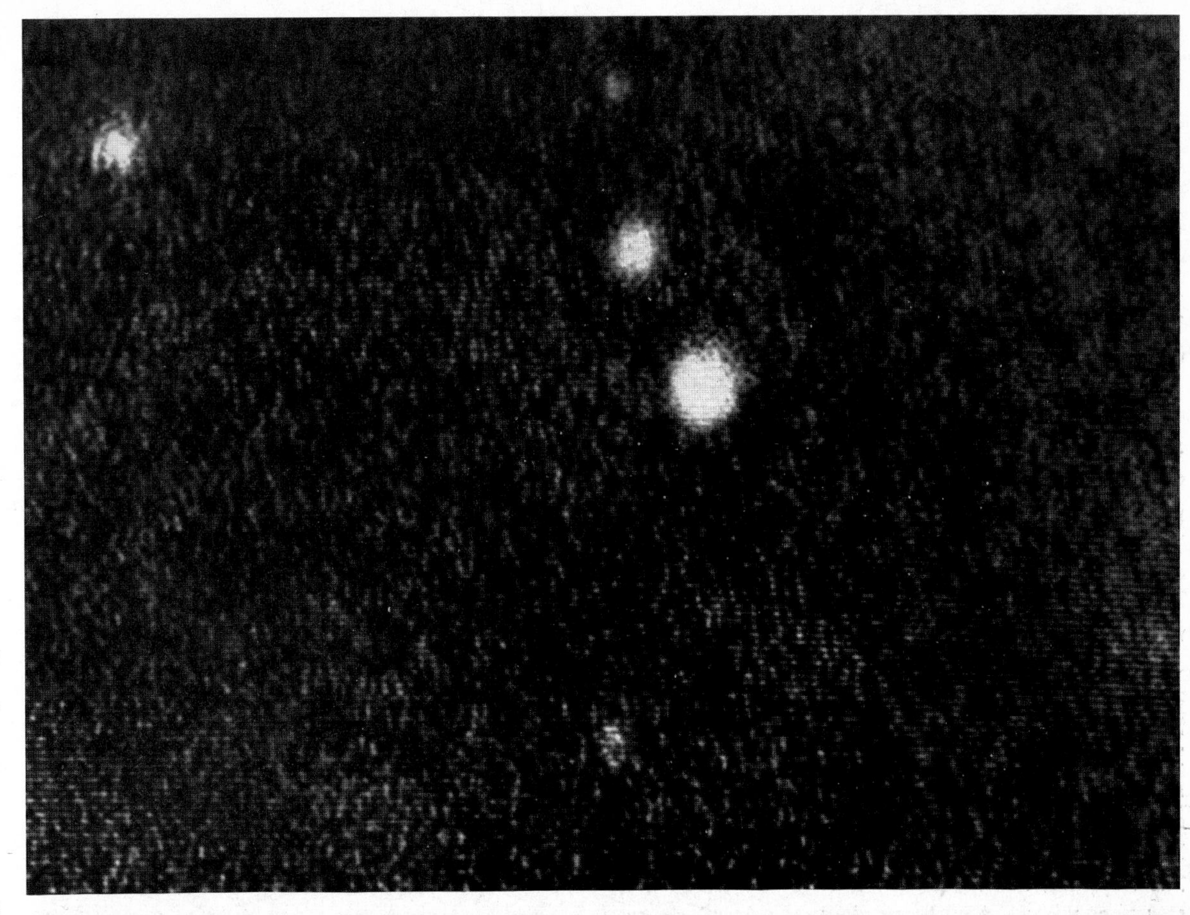

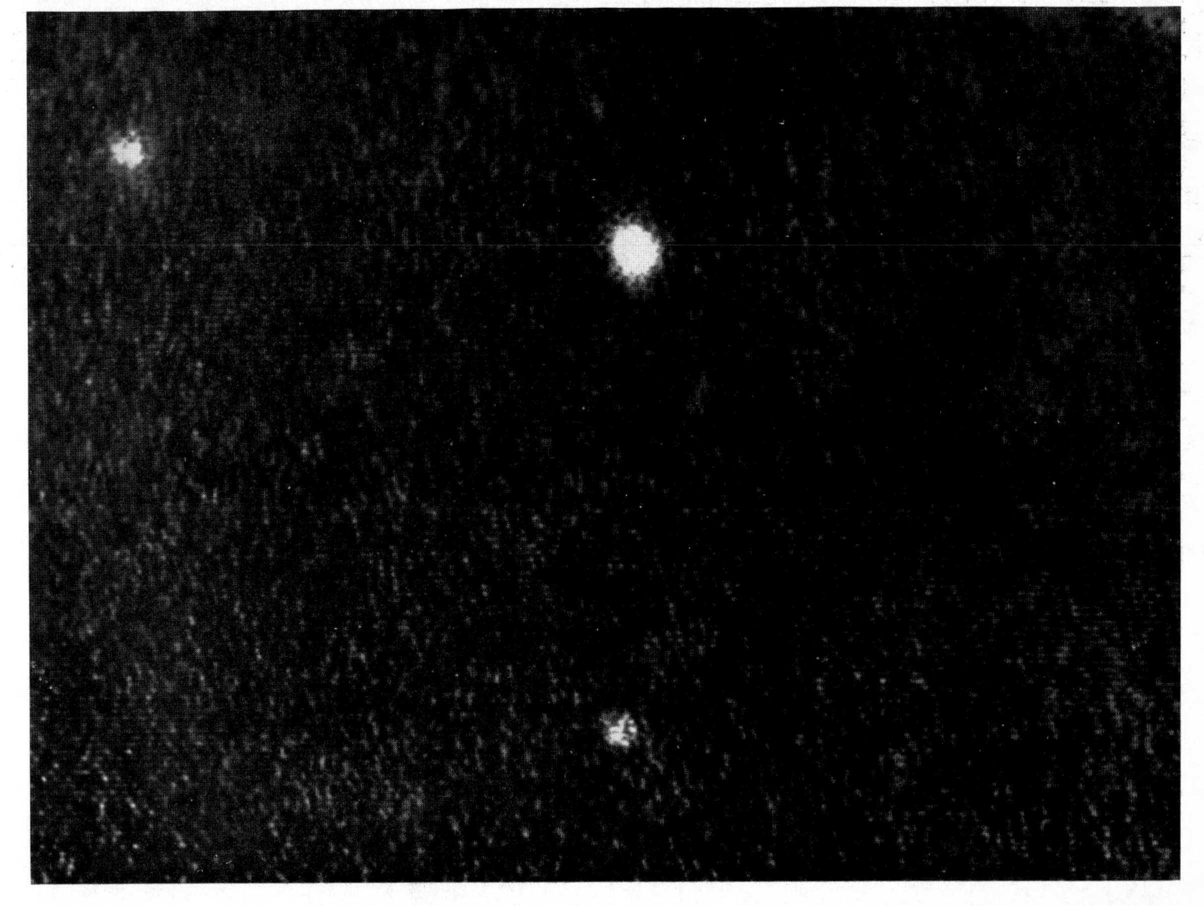

Hertzsprung-Russell diagram, they show, when linked together, the life story of a star. Since the numbers obtained by observations can also be included in the diagram, it is possible to compare them with theoretical calculations. The comparisons carried out indicate that the modern star models in some of the main phases of development are good approximations of the actual state of affairs. Above all, however, more work is still necessary on the very early and very late stages of development, such as the formation of the white dwarfs.

Disregarding a few details of little importance in the final result of stellar development, the future development of the Sun or related stars will be similar but significantly slower than that of the stars with a higher mass. From our knowledge of stellar evolution, it may logically be said that in a few thousand million years our Sun will inevitably develop into a giant star of reddish luminosity which will swallow up the inner parts of the solar system. Before this happens, the climate on the Earth will become so unbearably hot that our descendants will have to make arrangements in good time for a change of residence before the ground under their feet is burnt up.

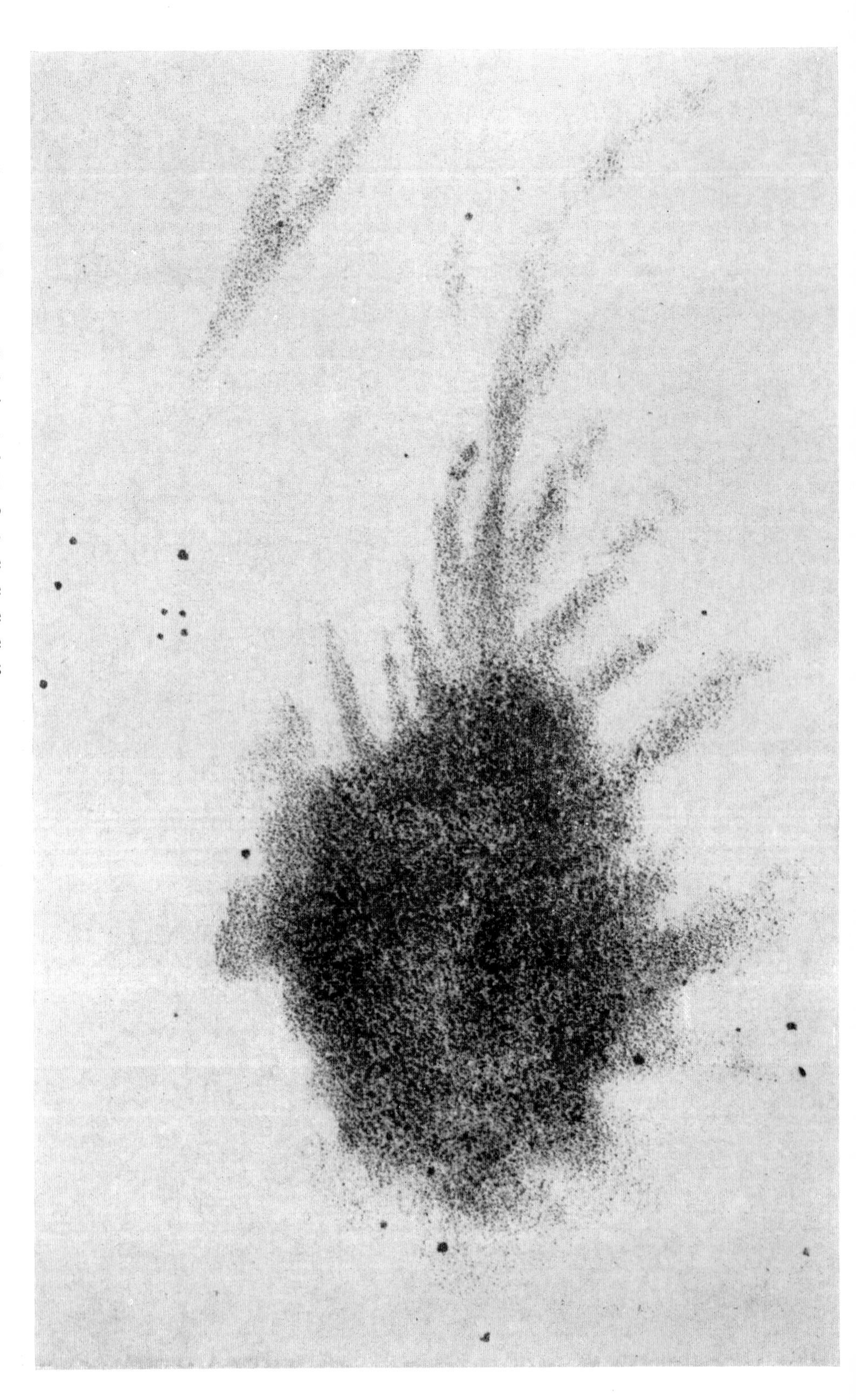

Chapter IV
Condensed matter in Space

Most of the matter in Space, as we have noted, consists of gas. For the most part, this gas is "organized" in stars, and only a little of it roams around through Space as interstellar gas. On a modest scale, however, the gas condenses and forms liquids and solids. Although it is estimated that only a few parts per thousand of cosmic matter are in the condensed state, the significance of this for the development and decline in the Universe is nevertheless remarkably great.

The planets orbiting the Sun are of solid and liquid matter and the innumerable small bodies of the solar system also consist of matter in the solid state. Most probably the planetary system of the star we know as the Sun is not an isolated case and does not represent a museum piece in the Milky Way system. There are indications now that other stars are also orbited by dark bodies of low mass. However, the present state of knowledge indicates that the emergence and development of living beings is limited to the surfaces and atmospheres of suitable planets. The study of condensed matter in Space thus includes the fascinating question of the existence of extraterrestrial life as well.

Solid matter in the shape of great clouds of tiny particles of dust in interstellar Space is another important form of the condensed state in the Cosmos. These inconspicuous but widely spread objects, which weaken and falsify light and thus complicate the work of astronomers, play a major role in the emergence and development of stars and of planetary systems.

The scope of modern research

While very little is yet known about cosmic dust, particularly its composition and origin, and almost nothing is known about other planetary systems, the investigation of the condensed matter in our cosmic homeland, the solar system, is already fairly advanced. This is because the telescopic observations

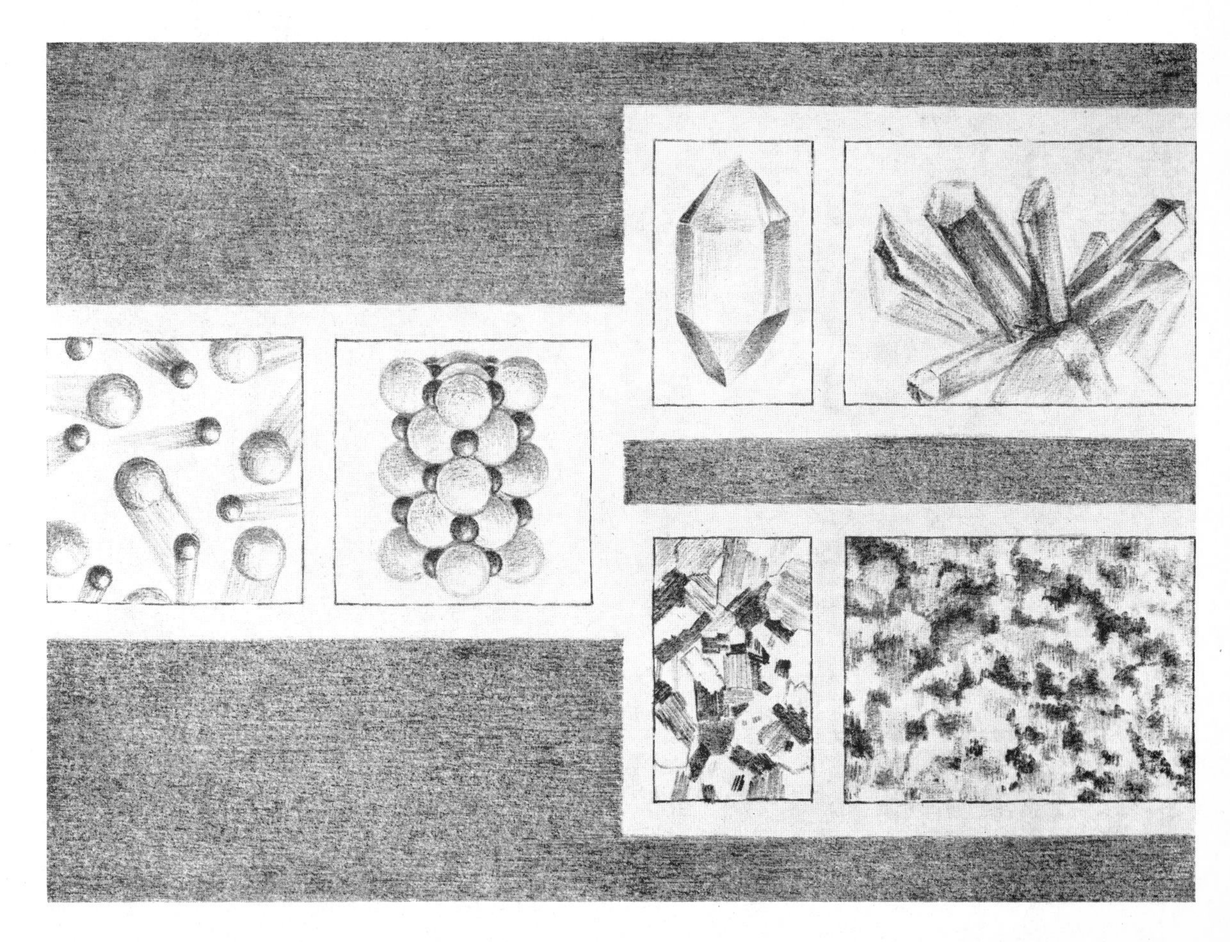

113 The atoms of a gas move around in haphazard fashion (left), but those of a solid body form a crystal lattice (center). When it can develop undisturbed, a solid body assumes the shape prescribed by the lattice structure (top center). Under natural conditions, impressive mineral crystals are formed (top right). When different substances crystallize at approximately the same time, they often lose their regular external appearance and take on a characteristic structure (bottom center). Under natural conditions, mixtures of minerals are found as rocks. The different componentscan sometimes be identified with the naked eye alone (bottom right).

114 Laboratory tests of very great astronomical importance include isotope counts and rock dating. The picture shows installations at the Isotope Geo-Chemical Laboratory of Freiberg Mining Academy used for carrying out these tests.

115a, b With the polarizing microscope, mineralogists can even reconstruct individual events in the history of the rocks investigated. On these two photographs of transparent sections lots of parallel lines, so-called deformation lamellae, can be seen in certain minerals. They are to be attributed to the momentary effect of extremely high pressures, the infallible sign that a meteorite has struck here. The left picture (a) shows a plagioclase crystal in the rock of the Nördlinger Ries, one of the largest terrestrial craters caused by a meteorite, while the right one (b) is a photograph of a pyroxene crystal in rock from the Ocean of Storms on the Moon.

and measurements typical of astronomy are very effectively supplemented by experimental methods, since some celestial bodies are within the range of Man and of his instruments. To the data obtained with telescopes there are added the findings from laboratory tests on rock samples from the celestial bodies in question and the field data acquired by on-the-spot inspection and measurements.

Telescopic observations, laboratory findings and field tests form the basis of modern research on the solar system, although the latter two were specifically developed for the investigation of the Earth by chemists, physicists, mineralogists and geologists after a great deal of painstaking detail work. When, at the beginning of the last century, it was realized that stones from Space occasionally fall on the Earth, there was no delay in using terrestrial methods of research for the examination of these messages from far away. These "celestial stones," the meteorites, are now the most thoroughly investigated form of condensed cosmic matter. This privileged status of the meteorites will be ended in the foreseeable future, however, by the achievements of space travel. For centuries, the Moon was at the focal point of telescopes but otherwise out of reach.

Nowadays, in the form of the rock samples brought back to Earth, the Moon is actually moving into the laboratories of the chemists, physicists and mineralogists, so to speak. Astronauts with a geological training and robots programmed for specific tasks are carrying out field investigations on the lunar surface. Automatic probes have begun the scientific exploration of Venus and Mars.

From the examination of terrestrial rock, meteorites and, very recently, of Moon rock, astronomers have obtained information which telescopes cannot give. They are learning, for instance, the proportions in which the various chemical elements are present in the structural material of the solar system, enabling them to draw general conclusions about the chemical composition of cosmic matter and about the origin of the elements in the Cosmos.

Furthermore, in the laboratory it is possible to ascertain the age of rock samples, providing astronomers with clues about the age of the solar system, for example. Finally, laboratory tests on rock samples and particularly field investigations can reveal isolated episodes in the history of the celestial bodies in question. Questions such as the circumstances leading to the "birth" of the planetary system of the Sun and of the origin of life on Earth then appear in a completely new light.

The blue planet

Naturally enough, it is the Earth, our home planet, of which we have the most extensive knowledge. Of the numerous facts available, those of most interest to astronomers refer only to the "astronomical" characteristics of this planet, enabling it to be compared with its sisters, the other planets. Such a comparison can provide the key to the understanding of these other celestial bodies. If it were possible to investigate the Earth from an astronomical viewpoint, as from an observatory situated in interplanetary Space, for instance, it would immediately be noticed that it has attributes of a special kind. It is the only planet of the solar system that has a bluish color and, through a telescope, it would be apparent that the blue planet is frequently covered by white clouds. Through the gaps in the constantly changing cloud, a surface would be glimpsed on which dark seas, light-colored continents and white polar zones would be revealed. In contrast to all the other planets, the atmosphere of this planet contains large quantities of oxygen; incidentally, this atmosphere causes its bluish appearance.

The Earth would also provide surprises for open-minded radio astronomers since the blue planet would emit radio noise of an extremely unnatural kind. Any intelligent living being capable of detecting such a noise would probably recognize it as a sign of a technically advanced civilization.

From a point fairly close to the blue planet, as shown by the numerous photographs taken on manned space flights, it is possible to make out a series of characteristic details which practically ask to be compared with other celestial bodies. Particularly striking is the extensive area of water. As the oceans, this water covers two-thirds of the Earth, and is the reason why the polar areas have a thick layer of ice. What is more, the oceans are linked up by a widespread network of waterways across the continents. The relief of the land area shows clear signs at many places of the erosion, shifting and deposition of material. The frequent occurrence of water and the weather processes taking place in the atmosphere provide an adequate explanation for the shaping of the Earth's surface by external forces.

However, at many points on the Earth there is evidence of the effects of internal forces such as, for example, the craters and lava fields of the volcanoes. Faults and dislocations reveal the presence of tensions in the Earth's crust which are released with formidable effect, but the clearest language is spoken by the mighty folds of the massifs which, as mountain chains, encircle the globe. In other words, the face of the blue planet is marked by scars and folds on the one hand and merciless erosion on the other.

Of interest also is the geophysical description of the Earth. Our planet has a magnetic field whose lines of force emerge from or disappear into the Earth near the poles of the axis of the Earth's rotation. This

116a, b The Moon is now the scene of field investigations. The left photograph (a) shows geological equipment on a rock in the Fra Mauro Highlands. The right photograph (b) records the geological inspection of the Hadley Rill.

magnetic field exerts a major influence on the physical processes in the area of Space around the Earth, beginning with the uppermost layers of the atmosphere which consist of highly rarified ionized gas, and extending to the zones of captive charged particles surrounding the Earth like a shell and forming the radiation belt discovered by the first artificial satellites in 1958.

The lower and denser layers of the atmosphere are almost an hermetic seal for the Earth, protecting it from cosmic influences which could endanger living beings. Like the water around the Earth, the envelope of air enclosing it has been formed by volatile components escaping from the interior of the Earth. A considerable change in its composition has taken place in the course of the Earth's history. Under the influence of vegetation, the original atmosphere containing no free oxygen, but rather reducing gases, has now become an oxidizing one.

The solid surface of the Earth is formed of rocks which mainly consist of mixtures of silicate minerals, with feldspars predominating. Depending on how they were formed, a distinction is drawn between the igneous rocks formed by the solidification of a

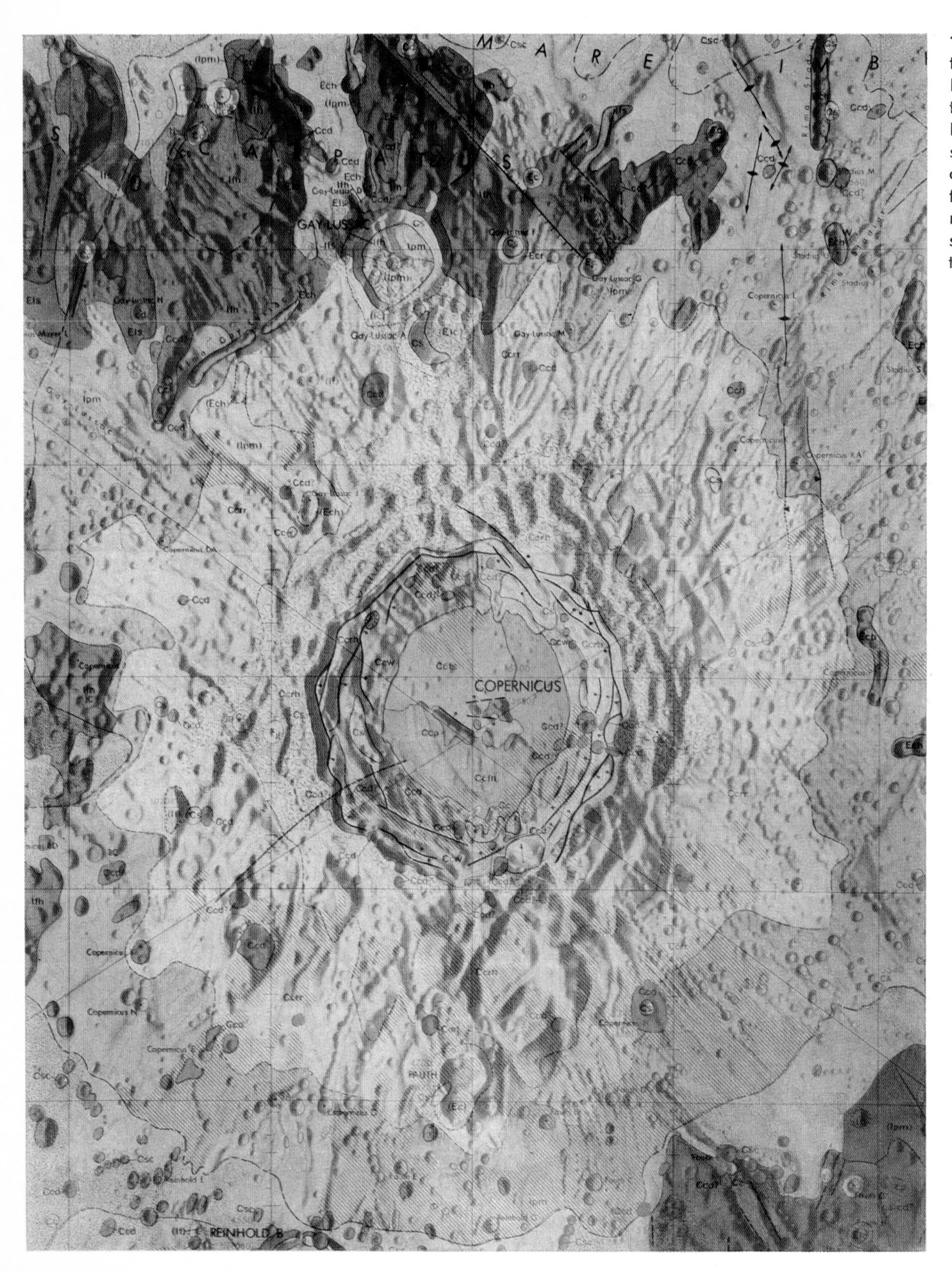

117 The first geological atlas of the Moon (scale 1:1,000,000) has been compiled from photographs taken by terrestrial telescopes and by cameras mounted on lunar satellites. The map shown is that of the Copernicus region. The different ages of the formations are marked by different colors. Other symbols are similar to those on terrestrial geological maps.

118 Collaboration between different disciplines is essential for the investigation of condensed cosmic matter. The darker the shading of the squares in the table, the larger is the extent of present-day knowledge about the various objects (Earth, Moon, Mars, Venus, Jupiter, meteorites, other planetary systems, cosmic dust). The telescopic studies of the astronomers are divided into measurements, observations in the basic sense of the word and spectroscopic studies. The laboratory tests carried out on rocks include chemical analyses, physical isotope counts and the petrographical investigations of the mineralogists. The field investigations are shown in the form of the actual geological field inspections and geophysical measurements, particularly seismological investigations. (The blank squares may be filled in by readers in due course!)

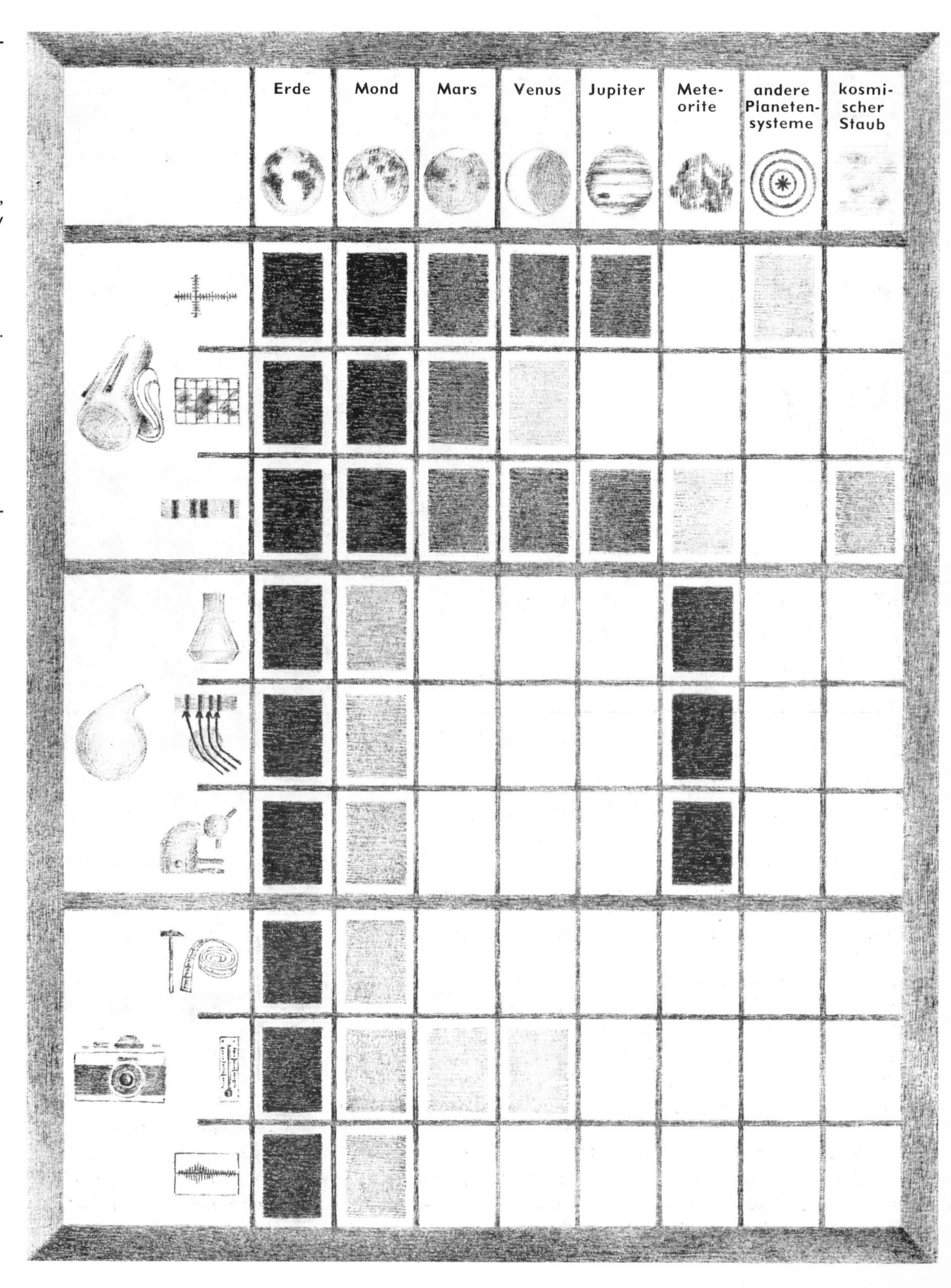

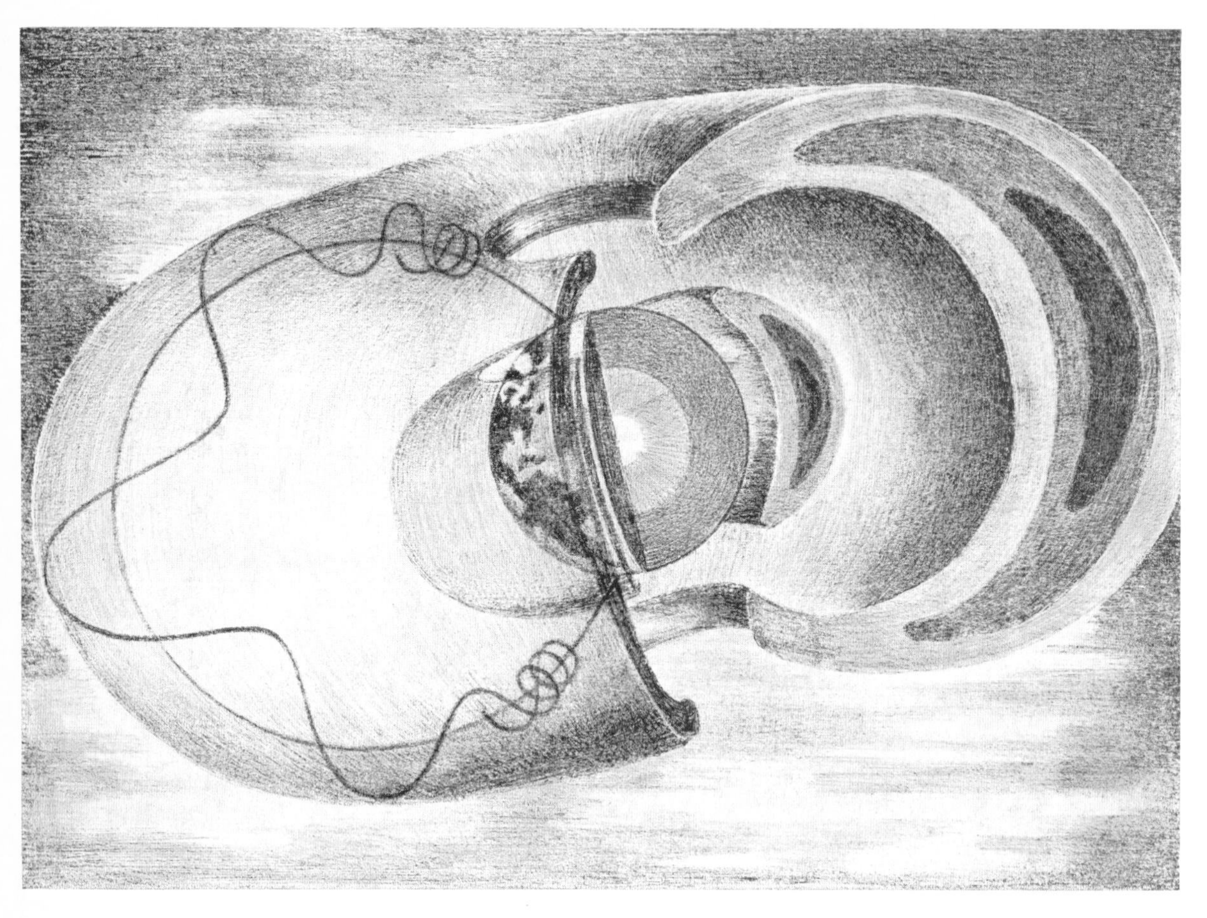

119 The Earth possesses a magnetic field which captures the electrons and protons coming from the Sun. These particles are thrown to and fro between the magnetic poles on spiral paths around the lines of force of this field. The left-hand part of the picture provides an idea of one of these paths. Electrons and protons of varying energy states collect in characteristic shells which surround the Earth as radiation belts. The magnetic field is probably caused by circulation movements in the electrically conductive core of the Earth which begins at a depth of about 2,900 kilometers (1,800 miles).

magmatic mass, sedimentary rocks which are the products of weathering deposited after transportation and turned into rock, and the metamorphous rocks, which are the result of the transformation of the first two kinds of rock by pressure and temperature. It is true that the Earth's crust, which is about 30 to 40 kilometers (19 to 25 miles) thick, consists largely of igneous and metamorphous rocks but three-quarters of the continental surface of the Earth is covered by sedimentary rocks which an observer, arriving on the Earth without previous knowledge, would consequently describe as the typical "Earth rock."

As indicated by the passage of earthquake waves through the Earth, the inaccessible interior of our planet is made up of a number of concentric shells with different physical characteristics. Underneath the Earth's crust and extending to a depth of 2,900 kilometers (1,800 miles), there is the mantle of the Earth, consisting of heavy rock. It encloses the liquid core of the Earth, the inner part of which apparently becomes solid again at a depth of 5,100 kilometers (about 3,200 miles) and probably consists largely of iron. The circulation processes maintained by the rotation of the Earth in its outer part are now considered to be the cause of the terrestrial magnetic field. It has been calculated by geophysicists that at the center of the Earth there is a pressure of several million atmospheres and a density of 15 to 20 grams per cubic centimeter.

Earth's unusual satellite

Let us conclude our examination of the blue planet with a look at its natural companion. Curiously enough, in relation to its primary, the companion of the Earth is the largest and most massive satellite. In this respect, the Earth's moon far outstrips the other thirty-one known moons in the planetary system. Many astronomers thus regard the Earth – Moon team as a double planet and, on the basis of various indications, assume that these two celestial bodies

to a large extent have a common past, and that there was a great deal of reciprocal influence in the course of their development.
The Moon, this unusual satellite of the Earth, has attracted the interest of men all through the ages. For most peoples, its spotty countenance is associated with numerous sagas and legends. In Antiquity, the light and dark regions were explained as areas of land and sea and it was also realized that it is always the same half of the Moon which faces the Earth and, from this, that the Moon has a spherical shape.
Through the invention of the telescope, astronomers became better acquainted with the surface of the celestial body closest to the Earth, and a new branch of astronomy, selenography, came into being. For the first time it was possible to study another world which was strange in many aspects and to compare it with the Earth. Lunar maps of ever greater accuracy were prepared and geographical names were given to selenographic details. Influential students of the Moon began to name the most striking features of the lunar landscape, the craters, for famous scholars.
Through a telescope it is apparent that the lunar seas are large and mostly clearly defined lowland plains. To all appearances, they have been formed by a massive flooding of the surface with lava. The dark, liquid rock, now identified as basalt, filled the depressions in the Moon's surface and solidified there. The "land" areas of the Moon, however, are lighter in color and higher. Occurring as coherent areas or as islands, they are probably the parts of the Moon's original crust which were not affected by the basalt flood. It is a surprising fact that the half of the Moon facing away from the Earth consists almost entirely of land areas.
Through the names given to the craters, the land areas on lunar maps look like a unique cemetery of scholars. In 1970, a commission of the International Astronomical Union also gave names to the great craters on the other side of the Moon which had been mapped in the meantime. Outstanding scholars of the past, who were forgotten or arbitrarily overlooked on the front side have been belatedly honored, while the great scientists of the Nineteenth and Twentieth Centuries have been put on an equal footing with their illustrious predecessors, and with spacemen and rocketry experts even being commemorated during their lifetime.
From its relief features, the land presents a much more varied picture than the sea. On the one hand, this is due to the craters, the round structural elements already mentioned, which are much more frequent on the continents than in the seas and often occur one within another, and, on the other hand, to the lunar mountains. Despite their terrestrial names —the Mare Imbrium, for instance, is flanked by the Alps, the Caucasus, the Apennines and the Carpathians—these mountains have nothing in common with the "fold" mountains of the Earth. They look more like upturned chunks of land or—as in the case of the mountains around the Mare Imbrium—like the edges of giant craters of the original surface before the appearance of the seas.
The round forms on the surface range in diameter from more than 200 kilometers (125 miles) down to a few inches and, as holes or flat depressions, are a major feature of the lunar landscape as it appears to terrestrial visitors. The term "crater" is ill-chosen since craters are nearly always associated with a volcanic origin. It is true that the great seas and lakes of solidified lava, their surface structures and other features which have been photographed by lunar satellites in recent years, are evidence that there must have been a great deal of volcanic activity on the Moon. Indeed, there are signs of modest outbursts of gas even today. However, most lunar geologists now consider that the majority of craters were caused by the impact of meteorites. In the century-long controversy between the supporters of volcanism and the proponents of the meteoric bombardment hypothesis it seems that common ground is being found. Many lunar specialists now agree that the relief of the Earth's satellite shows signs of forces emanating from the interior of the Moon as well as marks due to external influences from Space. It is feasible that the latter often played the role of a detonator for the action of forces from the depths. Due to the lower gravity of the Moon and the lack of an atmosphere, a meteorite striking the Moon generally does so with a far greater impact than would have been the case if it had hit the Earth. In addition, almost every event of this kind on the Moon is "preserved" for astronomical periods of time.
When a meteorite—and especially a small planet or a comet—collides with the Moon, it brings with it an enormous reserve of kinetic energy due to its high velocity. This is converted into heat while the "projectile" in question is driven into the lunar surface.

120 On photographs taken from points in Space, the Earth is enclosed in a bluish shimmer. The outlines of the continents, brown in color, can be seen between the dazzling white cloud formations with their interesting turbulence patterns.

121 a—d Depending on the silicic acid content of their minerals, the rocks of the Earth's crust are classified as acidic, intermediary, basic and ultrabasic. Four typical representatives are shown, the coarse-grained igneous rocks granite (a, top left), diorite (b, top right), gabbro (c, bottom left) and dunite (d, bottom right).

The resultant heat vaporizes both the material of the space missile and the rock in the direct vicinity of the impact area. The quantity of vapor released by this immediately attains a pressure of millions of atmospheres and blasts out a crater whose diameter is many times greater than the dimensions of the body striking the Moon. Rock turned into liquid glass is spattered around, the vapor liberated becomes condensed and, as a result, drops of glass rain from the sky. The shock wave spreading through the rock on impact ensures that the minerals in the debris thrown up assume characteristic deformations.

Incidentally, in order to study the effects of such events, it is not absolutely necessary to go to the Moon, since there are large impact-craters on the Earth. However, water and wind have "blurred" them so much that they are not always easily recognizable for what they are. Thus, the astronauts selected for the Apollo 14 expedition to the Moon were sent to the Nördlinger Ries in the south of the Federal Republic of Germany to study lunar geology. It was here, about 15 million years ago, that a small planet or a comet came down, causing a crater of 25 kilometers (15.5 miles) in diameter. The energy this represents would be equivalent to the simultaneous explosion of fifty of the biggest hydrogen bombs available today.

The most convincing arguments for bombardment by meteorites are to be found, of course, on the Moon itself. This is because the bedrock is covered by a layer of debris several meters thick. This consists of dust-like material which, under the action of the vacuum existing on the Moon (it does not have an atmosphere), has been baked together to some extent and contains pieces of rock of all sizes embedded in it. This moon dust is the result of a bombardment from Space lasting almost five thousand million years. It consists of fine mineral and rock fragments which frequently show signs of the high pressure stresses mentioned, and of tiny beads of glass which occasionally even contain traces of meteoritic material. The individual components do not necessarily originate from the bedrock, and transportation over considerable distances is likewise a consequence of the impact force.

While collisions of small meteorites with the Moon take place about as frequently as with the Earth, and long periods of time might pass between the arrivals of large projectiles, there is a contant stream of tiny micrometeorites measuring only fractions of a millimeter. On the Earth, these attract very little attention since the atmosphere completely slows down these particles even at great heights. The surface of the Moon, however, has been polished for thousands of millions of years by this cosmic "sandblast." Microscopically small impact impressions on the tiny balls of glass bear witness to this, as also do the round forms of certain Moon rocks which initially surprised the experts. Thus, despite the lack of an atmosphere, there is a kind of erosion on the Earth's satellite, a mechanical "weathering" without weather and, above all, without water.

Due to the successes of space travel, there has been a tremendous increase in our knowledge of the Moon since 1964. Several hundred pounds of Moon rock are now at the disposal of experts on the Earth. The expeditions dispatched to the surface of our natural satellite first carried out geological explorations. Seismometers today record tremors in the Moon's crust, and automatic measuring stations and mobile robots transmit a great deal of interesting data. There can be no doubt that most Moon research of the future will move away from the sphere of traditional astronomy. In the coming decades, geologists, geophysicists and mineralogists will have the chief say in the exploration of the Earth's satellite, but astronomers will take an extremely keen interest in the results of the investigations since they hope to get answers to fundamental astronomical questions such as how the Moon came into existence or how a planetary system originates.

Mars, the rust-colored planet

The most recent developments in space travel leave no doubt on one score: the Moon will not remain the only celestial body of the solar system to experience the "fate" of exploration by man. At the end of 1971 an intensive study of Mars began at very close range when, at almost the same time, the three probes—Mariner 9, Mars 2 and Mars 3—became artificial satellites of this planet. Mars 3 was even used to make an attempt to soft-land instruments on the surface of the planet. Mars, which is 150 times further away from us than the Moon when conditions are the most favorable, is most likely to be the next celestial body on which men will set foot.

The rust-colored planet, regarded since time immemorial as the greatest mischief-maker of them all, was Nergal, god of pestilence and death, to the Babylonians and was worshipped by the Greeks and

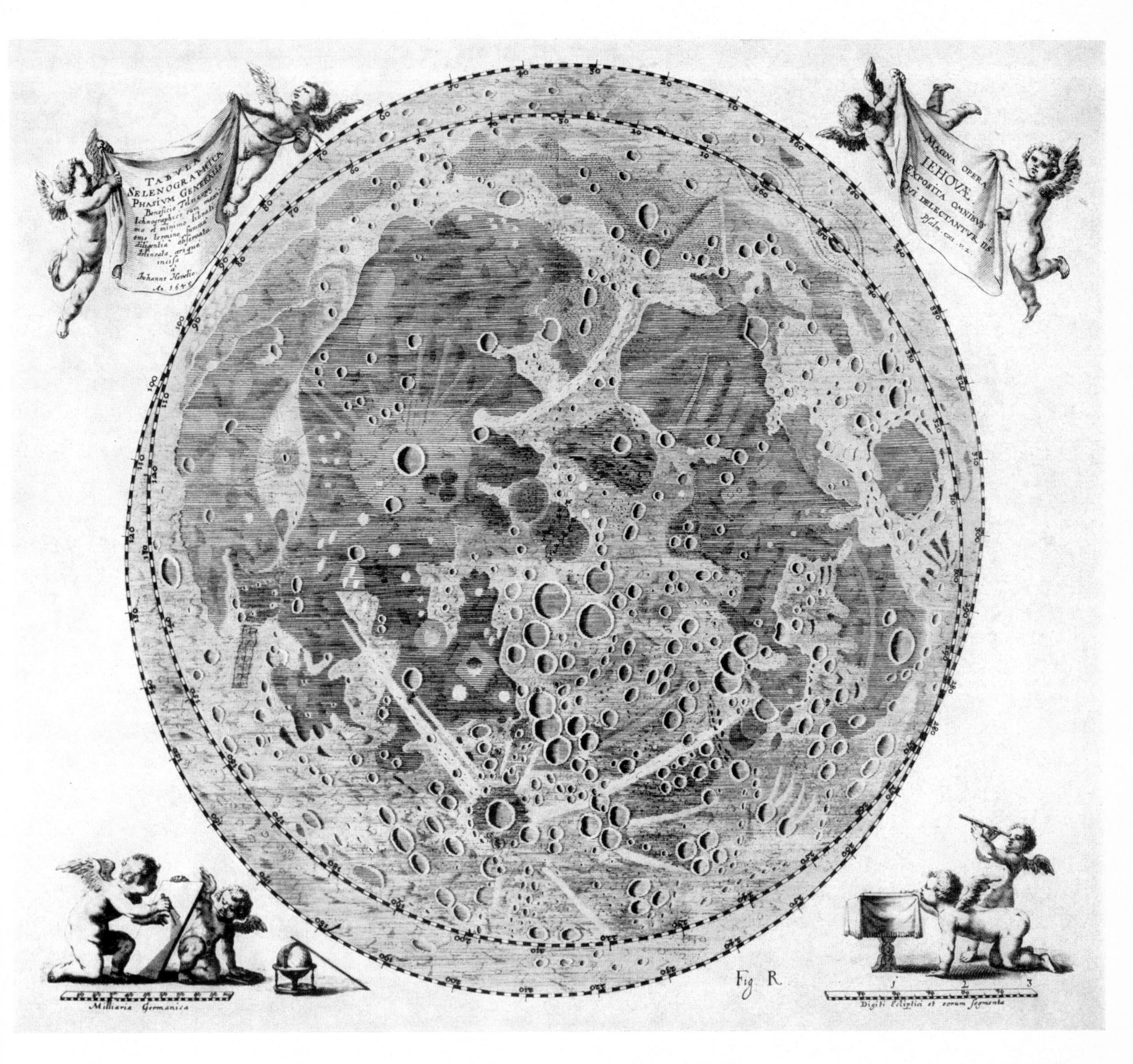

122 Following the invention of the telescope, a great deal of work was done in selenography in the Seventeenth Century. The most important work of this period was the *Selenographia* published in 1645 by Hevelius, from which this engraving is taken.

123 Characteristic of the Moon are the flat, dark areas of the seas, the lighter "mainland" regions and the craters. At the upper left of the picture, there is the eastern edge of the Mare Imbrium (Sea of Rains), bounded by the Apennines and the Caucasus. The first lunar probe landed between the three great craters of this region in 1959. Toward the east there is the almost circular Mare Serenitatis and, to the south of it, the Mare Tranquillitatis, at the southwest corner of which the first men from the Earth landed in 1969 on the Moon. In the lower part of the picture the beginning of the great southern continent of the Moon can be seen.

124a, b The various components of moon dust can be seen clearly on photographs taken with microscopes: fragments of fine- and coarse-grained basalts from the lunar seas, light-colored particles of anorthosite which probably represent the characteristic rock of the lunar mainland and particles of glass in the form of tiny balls, cylinders and irregular fragments (top). This moon dust is found as rock in lunar brecciae. The transparent section of a lunar breccia illustrated here shows the mineral and rock fragments and two pieces of glass embedded in a dark, glassy matrix (bottom).

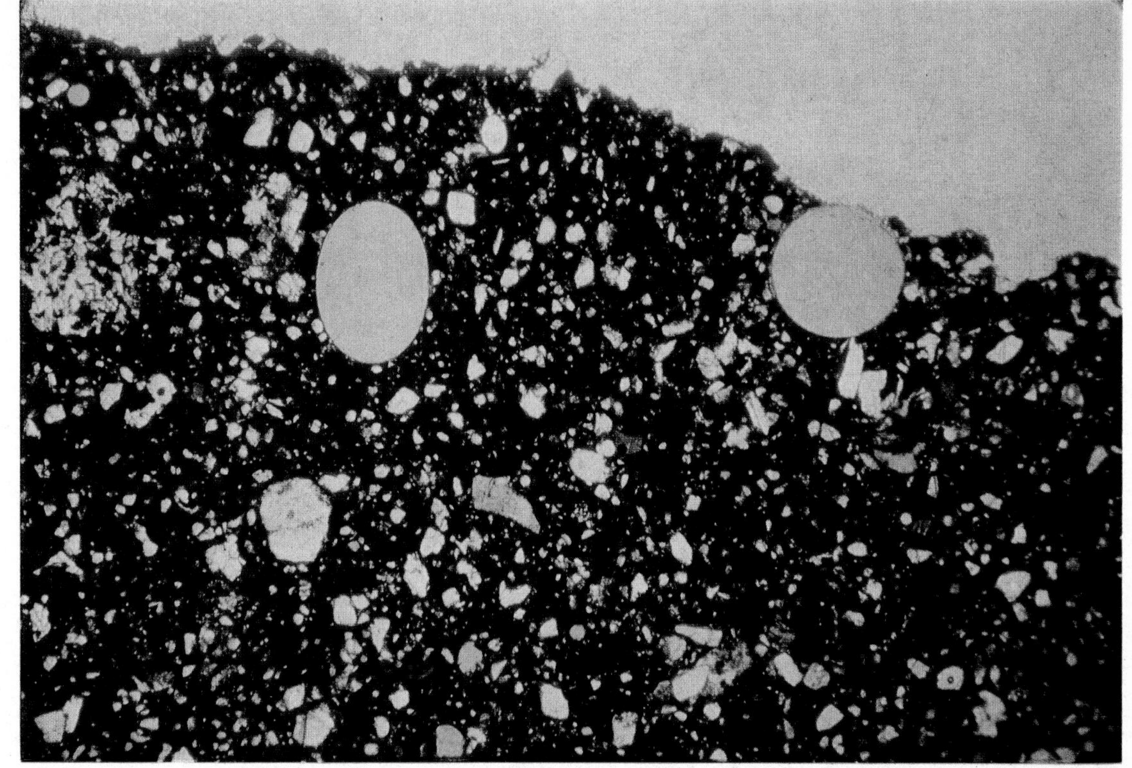

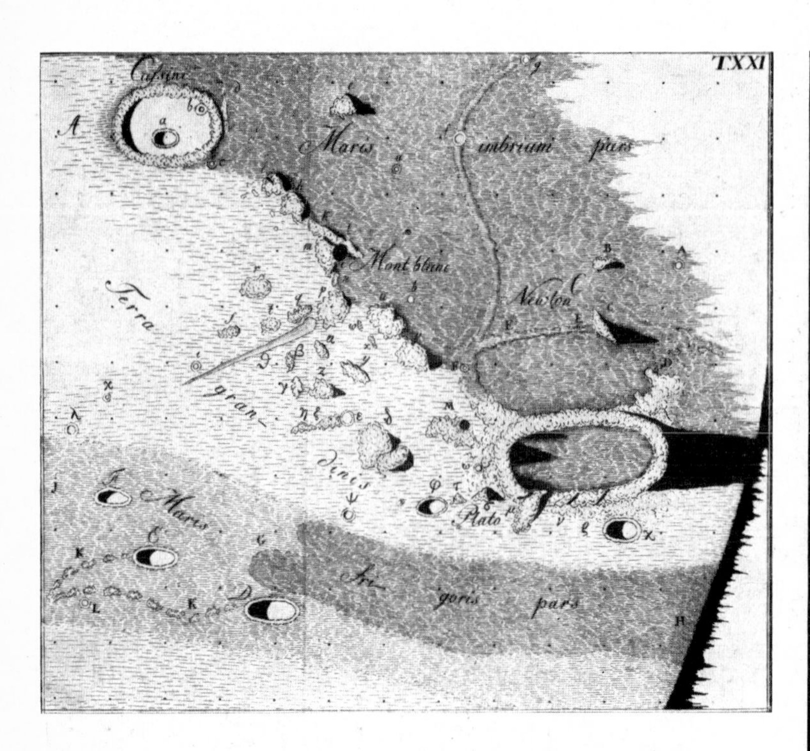
T.XXI
Maris
imbriani
pars
Mont blanc
Newton
Terra
gran-
dines
Plato
Ar-
goris
pars

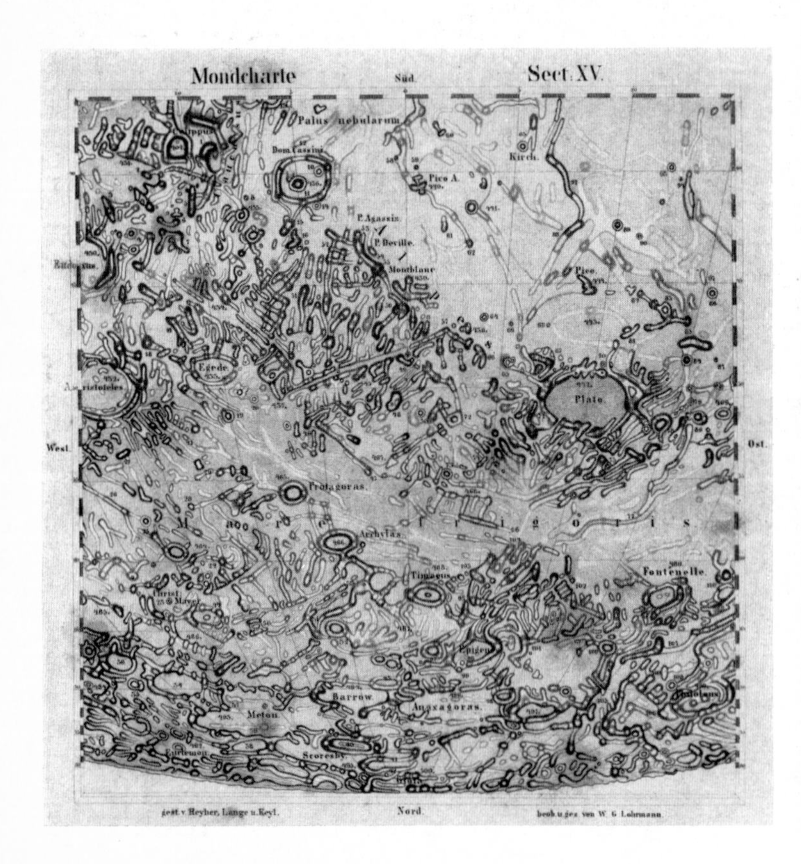
Mondcharte
Süd.
Sect. XV.
Palus nebularum
Plato
West.
Ost.
Protagoras
Archytas
Fontenelle
Barrow
Anaxagoras
Meton
Scoresby
Nord.

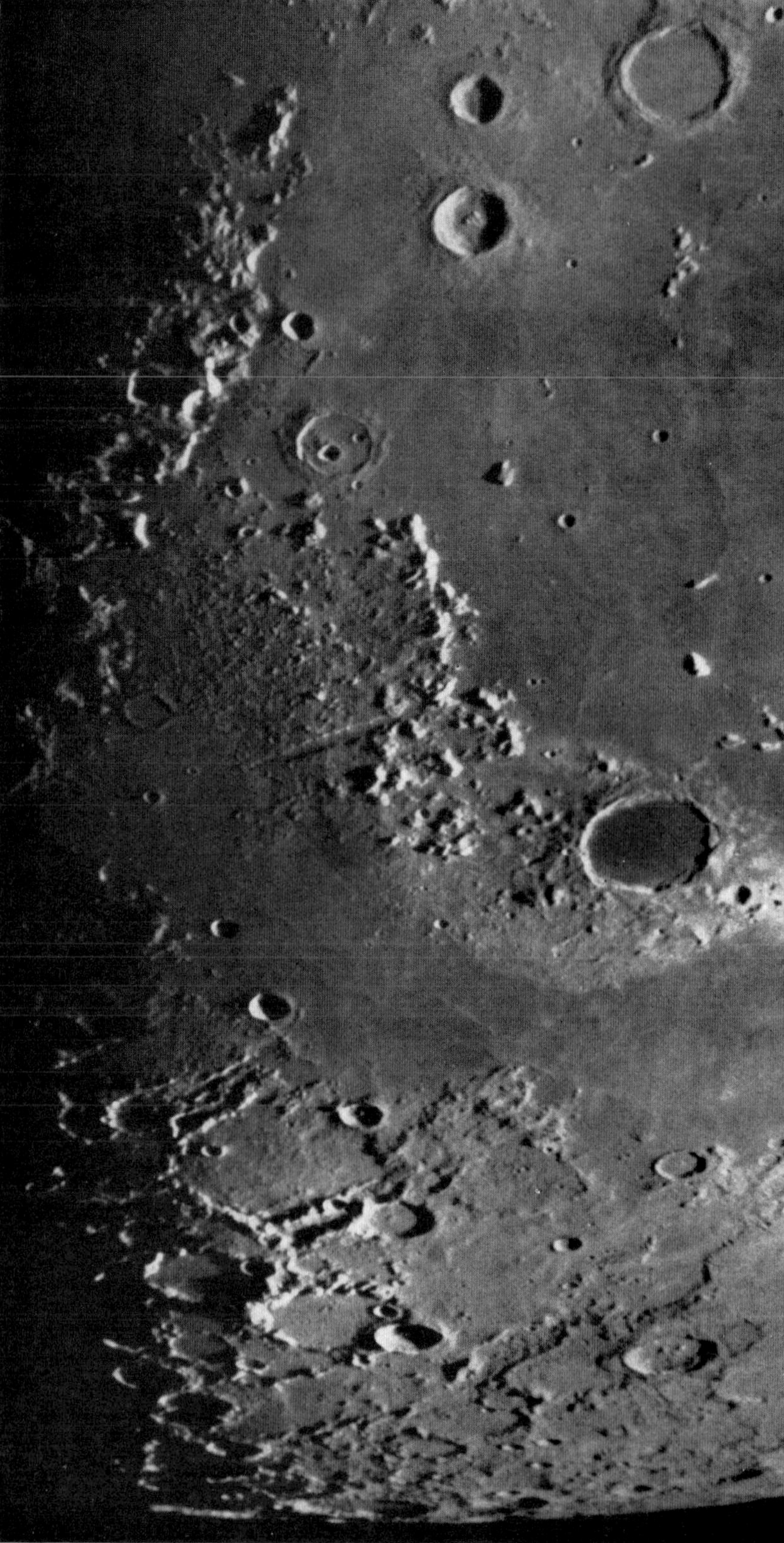

125a—d The development of selenography is apparent if pictures of the Moon's surface from different epochs are compared with each other. The four pictures show the lunar Alps with the famous Valley of the Alps (south is at the top of these pictures, in accordance with an old astronomical custom). The first illustration (a) shows this area as depicted in Johann Hieronymus Schroeter's *Selenotopographische Fragmente* of 1791. The next example (b) is a sheet from the well-known moon chart by Wilhelm Gotthelf Lohrmann of the Nineteenth Century. The photograph (c) is taken from the *Photographic Atlas of the Moon* published by Miyamoto in 1964. The last example (d) is a photograph taken by the moon satellite Lunar Orbiter 5 in 1967. On this picture, significant details can be recognized, including a rill that winds through the valley like a river.

Romans as the god of war. In recent times, however, it has attracted the special interest of astronomers. The observational material available concerning Mars enabled Johannes Kepler to work out the true shape of the planetary orbits. At the end of the last century, in particular, there was a rapid growth of interest in Mars on the part of many people. With the aid of improved telescopes, characteristics could be identified which invited comparison with the Earth and which had been sought in vain on the Moon.

Like the Earth, this planet revolves approximately once every 24 hours around its axis, whose inclination to the orbital plane is the cause of seasons similar to those of our own planet. Mars has white polar caps which clearly shrink in the Martian spring and on its surface light and dark patches are visible

126 Several automatic measuring stations are now at work on the lonely surface of the Moon. This group of instruments was installed by the Apollo 14 astronauts in the Fra Mauro area in 1971. It includes seismometers and instruments for active seismic experiments, instruments for detecting traces of a lunar atmosphere and movements of solar particles and a laser reflector.

127 At the foot of the lunar Apennines by the Hadley Rill, one of those winding valleys on the Moon which were probably "washed out" by low-viscosity lava, geological exploration trips were carried out with a lunar Rover by the Apollo 15 crew. In this picture, the rounded form of Mount Hadley in the background is particularly surprising.

128 Numerous photographs were taken by the Lunokhod 1 automatic moon vehicle which was remote controlled by a team of scientists and operated in the northern part of the Mare Imbrium, in the "Sinus Iridius," in 1970—71. This panorama view shows the nature of the lunar terrain, one fairly large and numerous tiny craters and the tracks left by the vehicle.

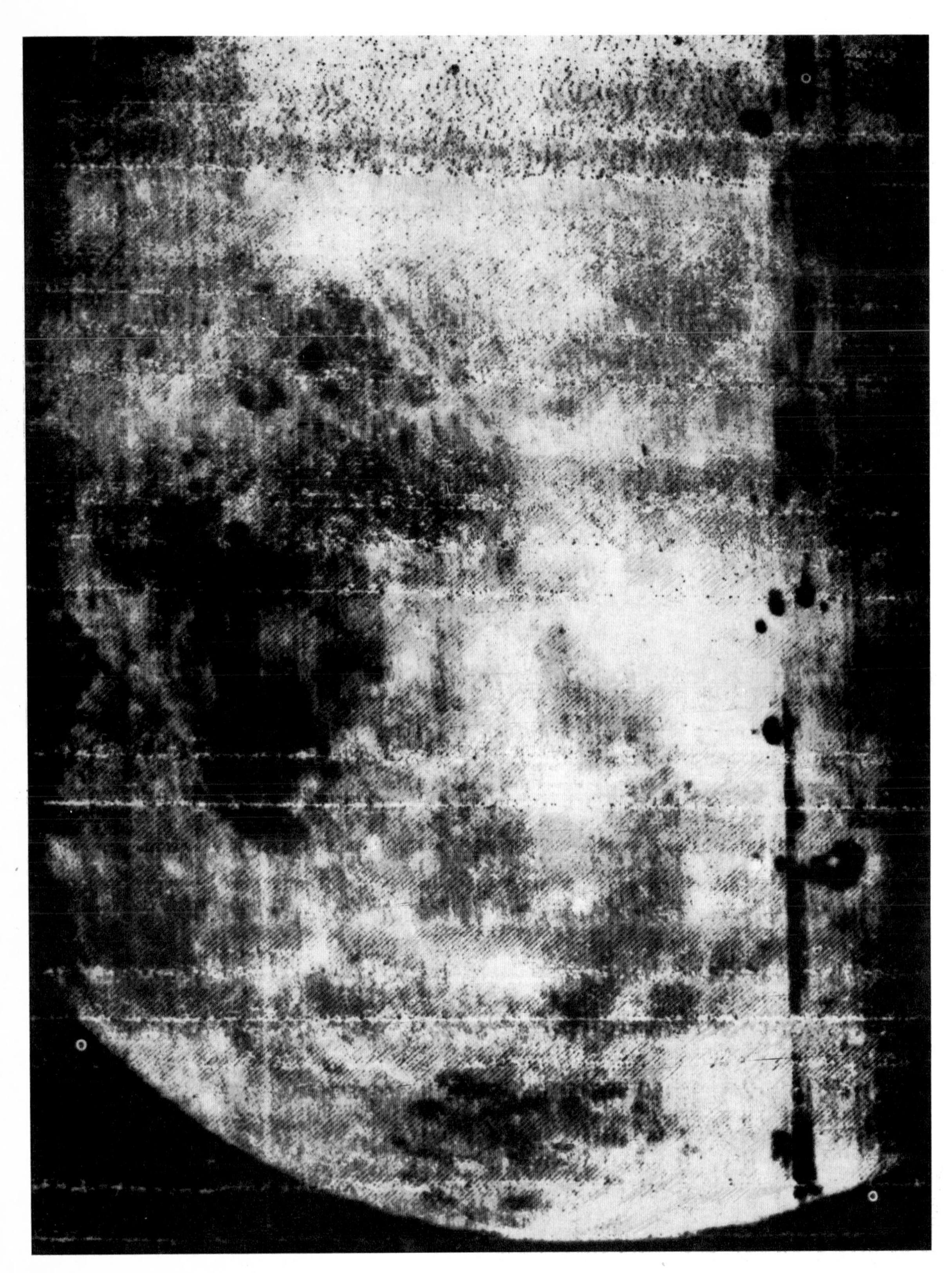

129a, b With a series of sensational pictures proving the continental character of the far side of the Moon, Luna 3 initiated the close-up exploration of the Moon's surface in 1959 (a). Within eight years, the far side of the Moon became almost as well known as the side that faces us. The second photograph (b), taken by Lunar Orbiter 3, shows the surroundings of the Ziolkovski crater which is 240 kilometers (150 miles) in diameter and which, with its extremely dark floor, can also be easily identified on the photograph taken by Luna 3.

which were interpreted, as on the Moon, as land and sea areas. In the last century, they were given names mainly taken from ancient geography and mythology. Expert observers noted clouds in the thin atmosphere of this interesting planet, and occasionally, even over large areas, the planet's details were obscured by sand or dust storms. The highest temperature reached on the ground at the Equator is about the same as that prevailing in the middle latitudes on the Earth.

When, in 1877, the Italian astronomer Giovanni Schiaparelli discovered a delicate network of thin lines on the planet's disk, Mars became the most popular object in the solar system. Schiaparelli was cautious enough to call them *canali*, meaning ditches or grooves, but his contemporaries immediately turned them into the now famous Martian "canals"—waterways presumably connecting the seas and flanked by strips of vegetation and towns. Later, when nobody believed any longer that the seas were areas of water, the shortage of water on Mars which had in the meantime become apparent was used to bolster up the canal hypothesis. In this way, so it was said, the Martians distributed the precious water obtained when the snow melted in the polar regions over the dry desert areas of their homeland. Within the last few decades, a vigorous discussion took place between those observers who really could see them as a network of thin lines and those who, despite the greatest care or perhaps precisely because of this, only perceived a blurred pattern of dark blotches and thus insinuated that their colleagues were the victims of an optical illusion.

The dark areas, the Martian seas, confronted the astronomers with many riddles. Since they are not seas at all, it follows that a reflection of the Sun was never observed in them. When it became apparent that their outlines and coloring changed with the seasons, most Mars researchers did not hesitate to interpret them as areas of vegetation. To sum up, it can be stated that in the last few decades it became accepted that there was a certain similarity between Mars and the Earth. People liked to compare the Martian landscape with the deserts in the highlands of Tibet—only the atmosphere of the red planet had to be imagined as considerably more rarified than that of the Tibetan high plateau.

When, in 1965, the television camera of Mariner 4 took the first close-up pictures of the surface of Mars from a distance of 10,000 kilometers (6,210 miles),

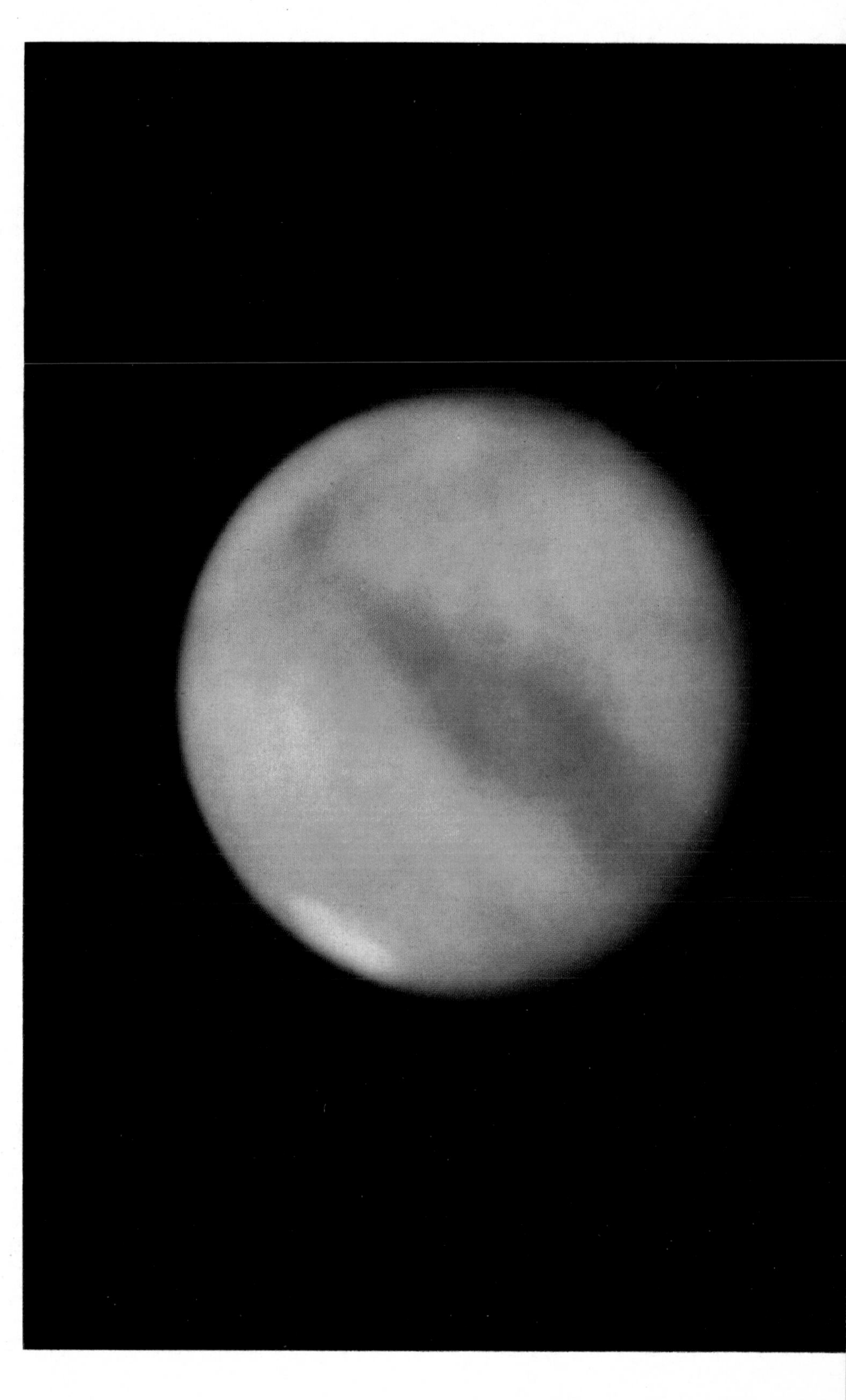

130 Bright and dark areas and a white polar cap can be seen on the surface of Mars, the rust-colored planet.

131 Compared with the Moon, the disk of Mars seems to be very tiny. The picture gives a good idea of the distance separating these two celestial bodies.

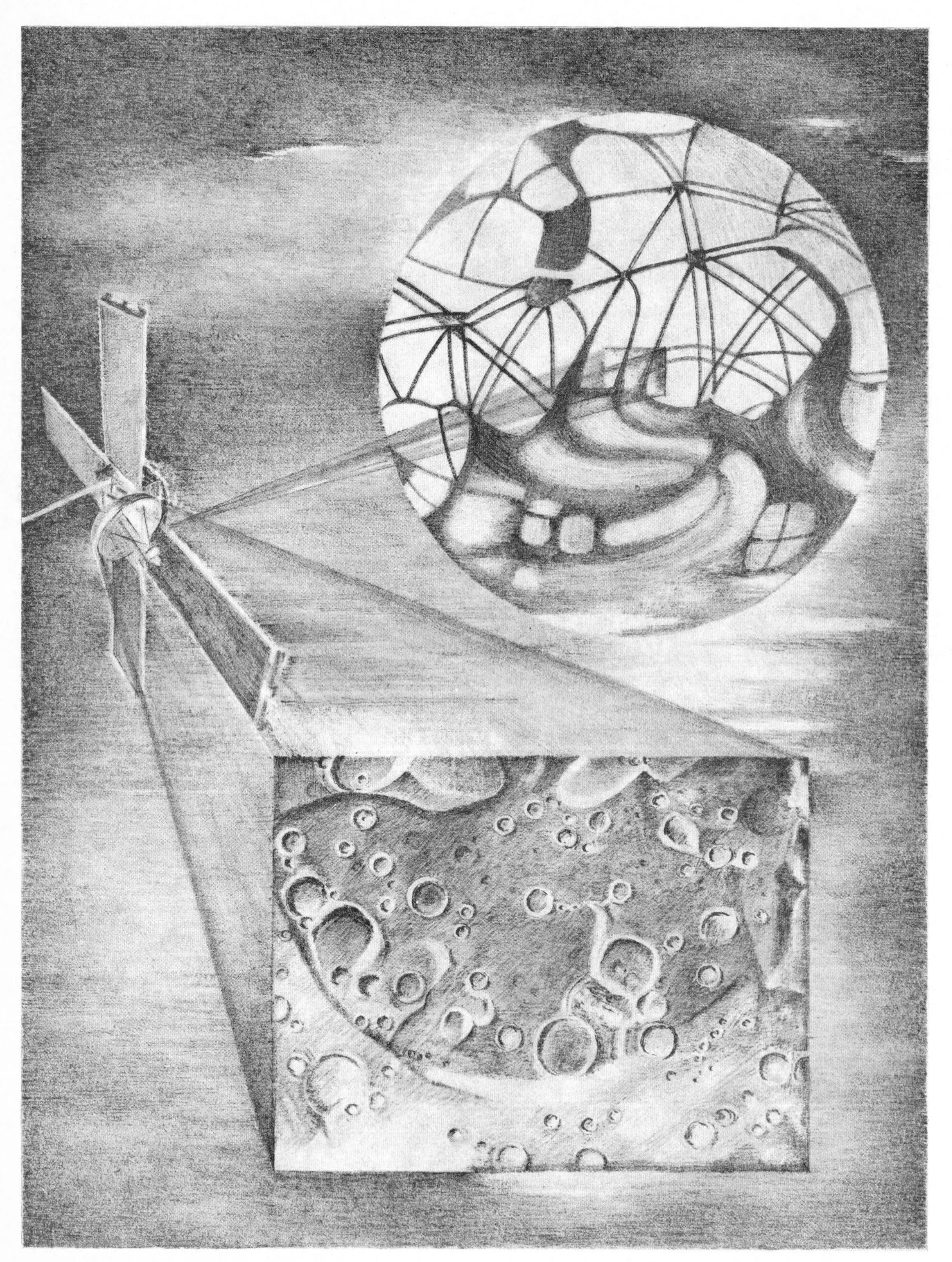

132 The cameras of space probes and artificial satellites are taking a closer look at the surface of Mars. In the past, the maps of Mars produced by prominent observers often showed the Martian canals as the dominating feature, but the new maps depict a crater landscape similar to that of the Moon in many respects.

the experts and many disillusioned Mars enthusiasts were astonished at the sight of numerous craters. Opinions about the much-discussed planet abruptly changed: Mars was not a second Earth, but possessed a typical lunar landscape. However, the pictures obtained in 1969 and 1971/72 make it clear that this standpoint is just as extreme as the previous one. On thousands of close-up photographs, covering the greater part of its surface, the rust-colored planet's own characteristic appearance can be seen. Apart from landscape elements similar to those on the Earth and on the Moon, it is marked by other features not found elsewhere until now.

The results achieved by the Mars satellites have fully confirmed the conclusions drawn about the relief of the planet based on investigations carried out with radar from the Earth. According to these, the maximum differences in height amount to about 15 kilometers (or 9 or 10 miles), which means that, when compared with its size, the surface of Mars is appreciably more rugged than that of our planet. Numerous examples to the contrary disprove the old idea that the dark regions are lowland areas. Since firm facts are now known about the extent of global dust storms, which has proved to be very much greater than was previously assumed, there is reason to believe that the seasonal changes in the dark regions are more likely associated with drifts caused by the wind than with a kind of vegetation. Drifting dust is also the probable reason why there are no craters to be seen in the great basin of the Hellas area, which is about 1,000 kilometers (or 600 miles) across, and other regions of the planet.

In addition to the numerous meteoritic craters, a series of subsidence craters of volcanic origin has also been discovered on Mars. These are known as calderas and are found together with other indications of geologically recent volcanic formations. The clouds occurring at certain points on the surface of Mars have long been known, and it may be that these are water vapor escaping from volcanoes. Even though on Mars there is no equivalent to the mighty mountain chains of our planet nor apparently to the great, drifting continental masses, recent findings nevertheless give the impression of much greater activity than on the Moon, for instance.

The riddle of the canals on Mars may be regarded, by and large, as solved. There is no overall network of straight surface structures linking the dark regions. Most of the canals are to be attributed to the inadequate resolving power of telescopes and only a few are associated with genuine topographical formations (faults, crater chains and so on). The "veil of myths" around the two tiny and mysterious Martian moons, discovered in the same year as the canals and since then the subject of much speculation, has likewise been removed. Television pictures show them as irregularly shaped celestial bodies of very dark material which are covered with impact-craters. If the surface of the planet is compared with that of its moons, the effect of erosion and other relief-shaping factors on Mars is strikingly apparent.

Measurements taken by the probes have also provided a more accurate picture of the atmosphere on Mars. The principal component of this thin "envelope of air" with a ground pressure of less than one percent of the terrestrial air-pressure seems to be carbon dioxide. Traces of water vapor were identified, but water in the liquid form cannot exist on the planet. However, it may be that a kind of permanently frozen ground exists there in which modest quantities of water are stored. The most recent photographs of Mars give proof of the fact that the past of the planet was not as dry as its present. While it must be assumed that the lowland areas were probably never covered by oceans, a whole series of clearly defined former "courses of rivers" can be distinguished, which, of course, are dry today. Moreover, there are indications that, in the past of this planet, the ground in the polar regions was subject to the grinding action of glaciers. The polar caps we can observe today consist mainly of a thin covering of carbon dioxide snow but they seem to contain, in their central parts, real snow, that is, frozen water.

Very little is known so far about the characteristics of Martian rock. From the visual features of the surface of Mars, various researchers have concluded that the rock there has a high silicic component. To a certain extent, these conjectures have been confirmed by the observations made of Martian dust very recently. The earlier idea that the color of the planet originates from the decomposition products of ferriferous minerals is also gaining ground. In this case, the rusty-brown appearance of the Martian continents really would have something to do with rust. Whether the God of War is, in actual fact, wearing a suit of rusty armor remains to be seen.

Venus, the planet of disappointment

Our second neighbor in the solar system is Venus, which moves within the orbit of the Earth. The radiance of this planet has probably impressed mankind since the dawn of history and, as the Morning or Evening Star, it has found a place in the literature of many peoples. The brightest star in the firmament after the Sun and the Moon, it was known to the Babylonians as Ishtar, the goddess of fertility, beauty and love. For a long time, however, astronomers could only subscribe to the high opinion expressed by this with respect to the aesthetic aspect, since for them Venus was the planet of disappointment.

After Galileo had discovered the phases of this planet at the beginning of the Seventeenth Century, generations of astronomers expectantly directed their telescopes at Venus time and again, but they were always disappointed since their searching looks sank in an ocean of cloud and fog. Dark spots on the disk of Venus were sighted from time to time, but the clouds did not move away. This is the reason why, until about a decade ago, it was not possible to determine the rotational period of Venus.

The veiled countenance of the goddess was naturally cause enough for numerous wild hypotheses concerning her true appearance. Since, in terms of size and mass, Venus might be termed the twin sister of the Earth, the planet was described in concepts of a markedly terrestrial nature. Thus, various scholars suspected that the veil of clouds concealed a mysterious tropical world with vigorous flora and fauna, while others thought of marshy forests of ferns and horsetails such as existed on Earth during the Carboniferous Period. Many also imagined that the planet was completely covered by water, while others visualized oceans of oil.

All these ideas are now outmoded, for it is clear that the surface of this planet is very hot. This was noticed for the first time in 1956 by radio astronomers who came across unexpectedly intense radiation from Venus in the microwave range, leading them to infer that the ground temperature must be extremely high. It should be pointed out that microwaves, unlike light and infrared radiation, can penetrate a cloud cover and can consequently supply information about the surface which emits them.

These conclusions were confirmed in 1962 when the space probe Mariner 2 carried out a close-up study of Venus, the data obtained indicating a ground temperature of about 425 °C. In contrast to this,

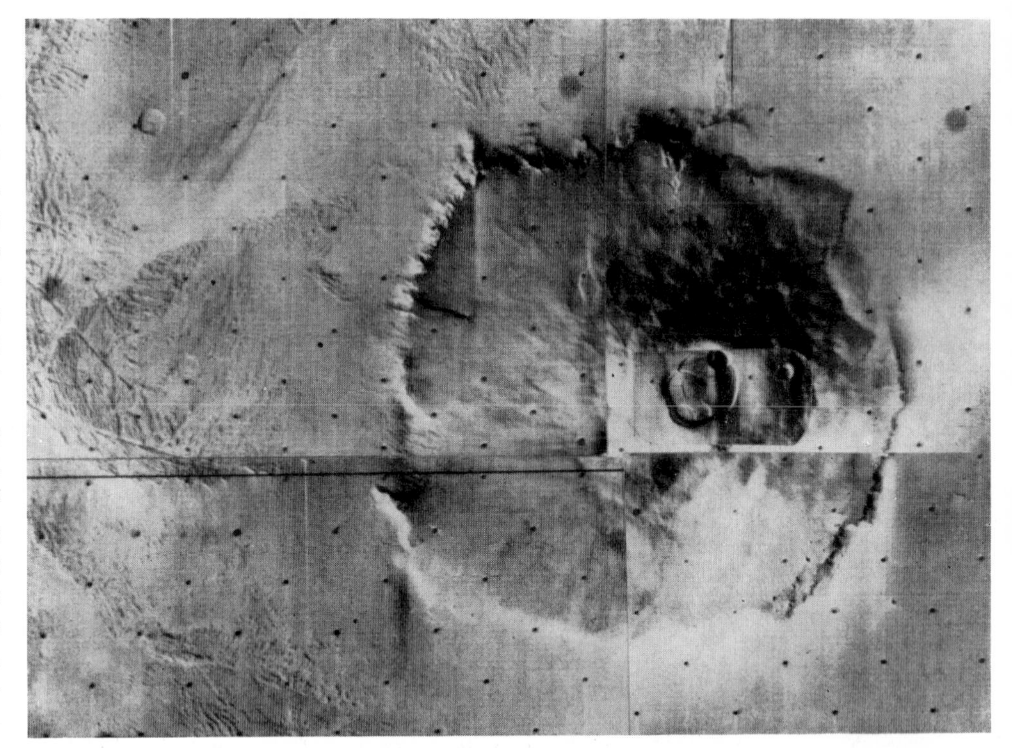

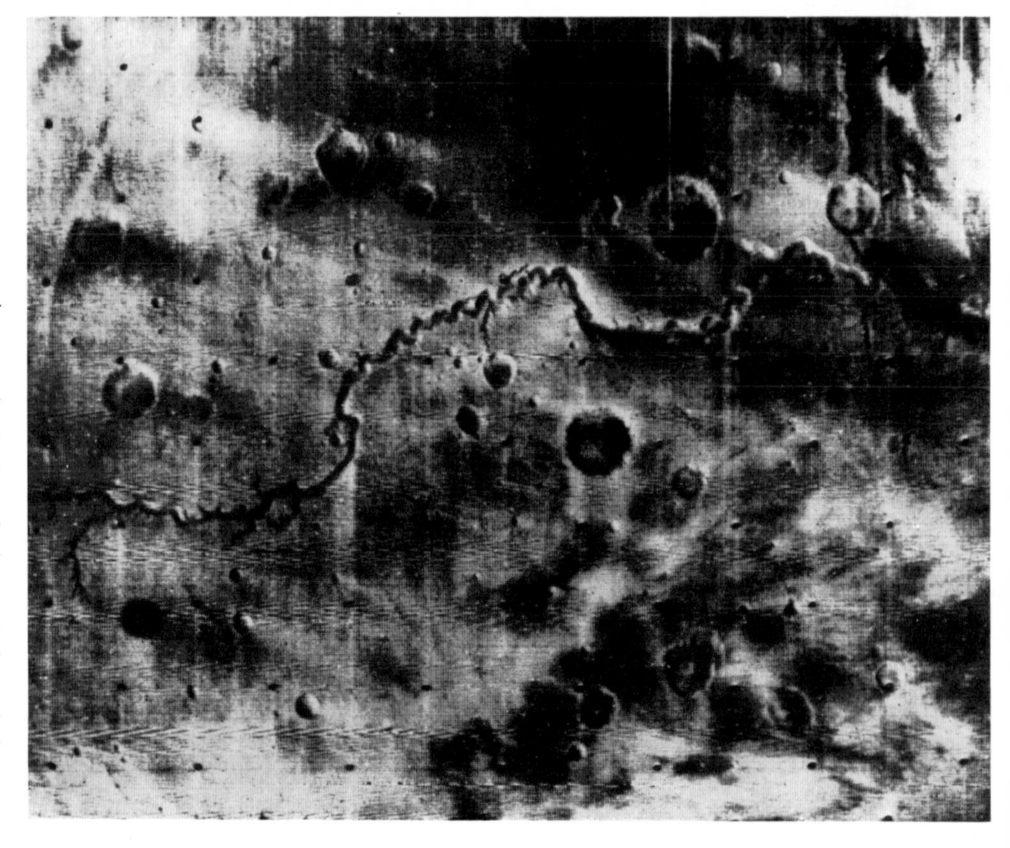

133a—c Numerous formations on its surface indicate that Mars is a planet of extraordinary geological surprises. The Nix Olympica proved to be a volcano massif which is nearly double the size of the greatest terrestrial formation of this kind (Hawaii). On a base of about 500 km (300 miles) in diameter and which steeply stands out against the surrounding plain, the mighty cone of a volcano rises to nearly 25 kilometers above the average level of Mars. Its slopes are covered with tongues and channels of lava. On its blunt peak, a caldera of 65 kilometers in diameter can be seen (a). A trench of about 4,000 kilometers in length follows the equator of Mars. The picture (e) shows a part of this trench (north is to the right) which is situated in the Tithonius Lacus region. The dotted line marks an area for which height data have been obtained. In appearance, it resembles a canyon and is about 100 kilometers (62 miles) wide and 6,000 meters deep (lower white arrow). Parallel to the main trench there is a chain of volcanic craters which probably follow the course of a volcanic fissure (upper white arrow). The valley (in illustration b) resembling the course of a river in appearance, is about 400 kilometers long and meanders through the landscape of the Mare Erythraeum. It is not at all unlikely that it is a real river valley.

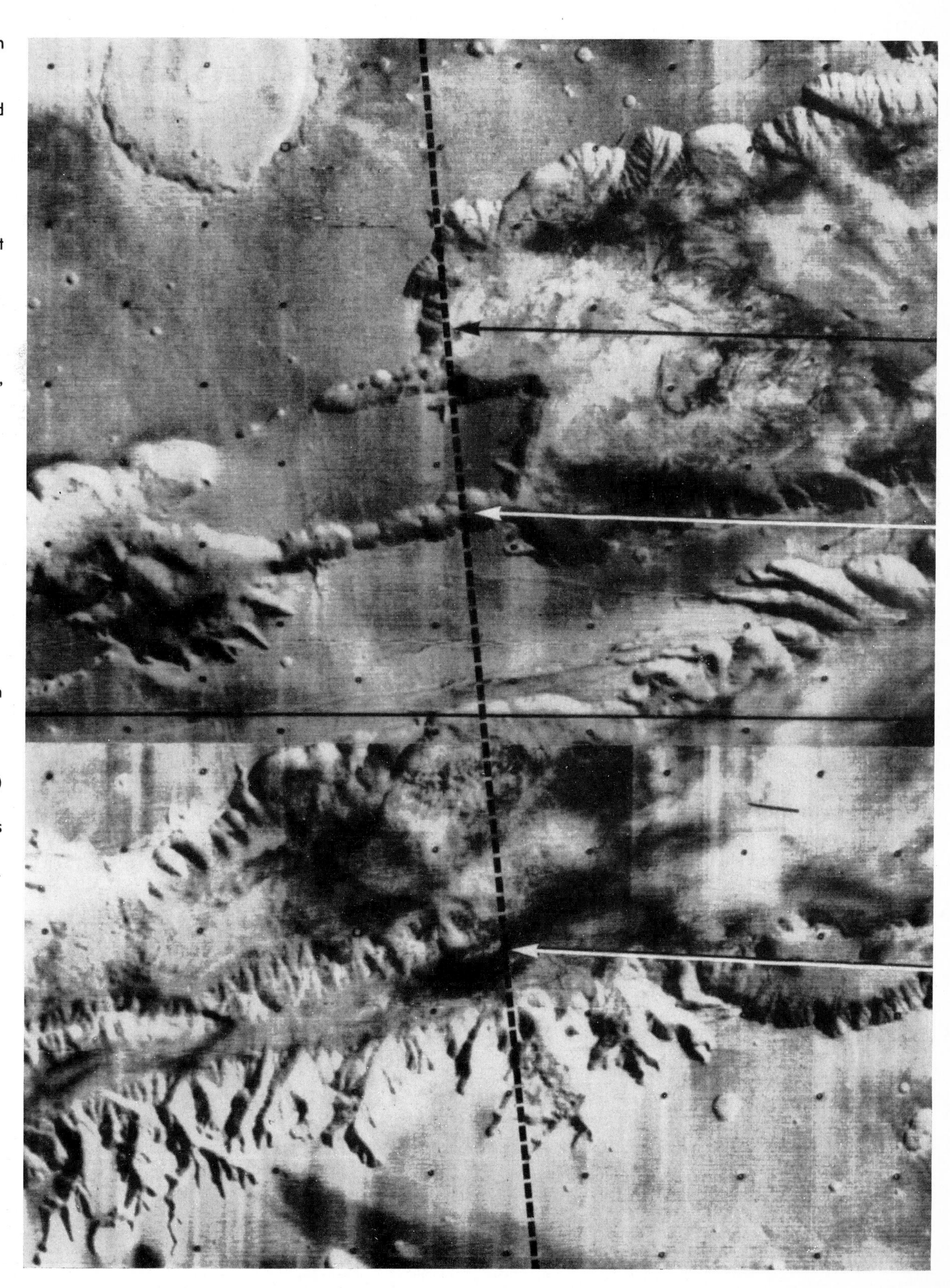

measurements of infrared radiation supplied a figure of only −40 °C, but this refers to the top of the cloud cover. Astronomers were presented with the fascinating picture of a planet glowing dark red and enclosed in a mantle of ice clouds.

Between 1967 and 1970, the probes Venus 4, 5 and 6 succeeded in bringing instruments into the lower atmosphere of the mysterious planet, and those from Venus 7 were even able to carry out brief measurements on the glowing surface of the planet before radio transmission ceased. The results obtained confirm that Venus is a planet of extremes. Although all the probes landed on the night side, the ground temperatures measured or estimated were between 400 and 530 °C, while the atmospheric pressure was between 60 and 140 atmospheres. The differing figures in the individual landing areas are probably caused by differences in height. The chemical analyses carried out during the flights through the atmosphere have confirmed the qualitative conclusions drawn previously from the spectrum of Venus. About 95 percent of "Venusian air" consists of carbon dioxide, the rest being made up of nitrogen and inert gases. Slight quantities of oxygen and water vapor have also been detected.

Incidentally, the composition of the lower layers of Venusian atmosphere and the clouds, which would appear to consist of ice crystals, and which are situated in the higher layers, explain the high ground temperatures. This is because the carbon dioxide and the ice absorb the infrared radiation emitted by the planet's surface in order to get rid of the heat which the ground of Venus has received in the form of light. Since this heat is thus prevented from escaping into Space, there is a build-up of heat under the cloud cover. Only when the surface has become sufficiently hot can the longer microwaves convey enough heat through the cloud cover so that a balance is obtained between the radiation from the Sun and the energy released through microwaves.

In short, the goddess is beginning to lift her veil, and what has been revealed up to now has little in common with that suspected in the past. What is appearing is not a landscape flourishing with life but a hot desert of sand in which no life can exist and which is probably cursed by storms. However, the last word has not yet been spoken about any of these aspects, and Venus researchers say that the reluctant goddess may admit many other characteristics during the scientific cross-examination in the coming decades.

134 The Martian moons Phobos and Deimos are rugged, crater-covered chunks of rock. All the speculation about the artificial origin of these celestial bodies is shown to be groundless by this picture of Phobos.

Stepsisters of the Earth

So far, we have considered the celestial bodies of the solar system to which astronomers have assigned the term "terrestrial" or "Earth-like" since they can be formally classified by a list of external features also possessed by the Earth. One of the most significant characteristics here seems to be the mean density since it can provide important information about the material composition. The terrestrial planets—Venus, the Earth, Mars and the Sun's closest neighbor, Mercury (of whose surface very little is yet known)—have fairly high mean densities. Those of Mercury, Venus and the Earth are between 5 and 6 grams per cubic centimeter and demonstrate that there must be heavy cores under the outer shells of rock. These are not possessed by the Moon with a mean density of 3.3 grams per cubic centimeter nor, in all likelihood, by Mars with just under 4 grams per cubic centimeter.

It is worth noting that these relatively dense celestial bodies inhabit the inner regions of the solar system, i.e., between 80 and 300 times the radius of the Sun. Furthermore, they orbit the Sun at characteristic intervals which may even be represented in the form of a mathematical expression. Johannes Kepler was the first who attempted to find such a relation. While engaged in this, he noticed the unusually large gap between Mars and Jupiter. Later Johann Daniel Titius (1729—1796) and Johann Elert Bode (1747—1826) publicized a mathematical law that suggested there was a planet "missing" after Mars, which they hypothetically inserted in the gap to complete the logical pattern.

On the far side of the gap, at intervals of 1,000 to 8,000 solar radii, there then follow five other planets, four of which—Jupiter, Saturn, Uranus and Neptune—have little affinity with the Earth in terms of their external features and could be called the "stepsisters" of the Earth. Since they have considerably larger dimensions and masses than the Earth-like planets, they are called giant planets and are known as "Jovian" or "Jupiter-like." Pluto, the outermost planet of the solar system, does not share these characteristics, so it was occasionally called Earth-like. For the time being, however, caution is advisable in classifying this planet which was only discovered in 1930.

A particularly striking feature of the Jupiter-like planets is their low mean density which ranges between 0.7 grams per cubic centimeter and 1.6 grams per cubic centimeter. The only chemical element which, in the condensed state, is compatible with such a low density at the pressures prevailing inside such great masses is hydrogen, and this must consequently be the principal component of the giant planets. It seems that the chemical composition of the Jupiter-like planets reflect the make-up of interstellar gas, the original structural material of the solar system. Due to their great masses and their vast distances from the Sun, these bodies were able to retain the highly volatile components hydrogen and helium lost to Space by the Earth and its sisters at an early stage in their history. The Jupiter-like planets may be compared, as it were, to juicy grapes while the Earth-like ones would be represented by raisins.

The stepsisters of the Earth display an unusual degree of activity, as can be easily confirmed by any observer. These great masses are spun around their axes in only ten to fifteen hours, and their vicinity is practically swarming with moons, the brightest of which were discovered as long ago as the Seventeenth Century. Each of these planets appears to have a small-scale solar system. The mighty atmospheres of the Jupiter-like planets are also characterized by incredible activity. It is not known how far below the dense cloud cover the surface of these planets is to be found. In addition to hydrogen and helium, the chief components of the atmosphere are methane, ammonia and probably water. A system of mighty winds maintains the striped appearance of Jupiter and Saturn. The famous Great Red Spot in the atmosphere of Jupiter, covering an area the size of the Earth, seems to be a vast whirlpool which, as observations show, exercises an attraction on the surrounding cloud systems. How this whirlpool remains practically stationary and what causes it is still unknown.

Theoretical research indicates that the interior of Jupiter consists mainly of solid hydrogen. At a pressure of more than a million atmospheres, it is changed to a metallic form, so that Jupiter, like the Earth, possesses a core of good electrical conductivity. Over the solid hydrogen there is probably an ocean of liquid hydrogen with all the possible condensates of hydrogen compounds floating in it as "icebergs." It also may be that this material is more like a morass in a swamp. Jupiter's metallic core is the probable reason for a powerful magnetic field the strength of which near the planet's surface can be expected to be ten times as large as that near the Earth's surface as

135 The analysis of radar waves reflected by the surface of Venus provided the first information about Venusian topography. It is possible that some of the dark zones, which reflect the radar waves particularly well, are areas covered with debris. Radar astronomy also succeeded in solving the riddle of the rotation of Venus. The planet completes one revolution every 243 days and, in contrast to all other Earth-like planets, turns in a clockwise direction when seen from the north pole (retrograde).

the recently obtained close-up measurements by the space probe Pioneer 10 show. The effects exerted by this strong magnetic field cannot be ignored. Jupiter is the most important radio source in the meter-wave range and a major one in the microwave area. The radio radiation emanates from a mighty belt of charged particles captured by the planet in the same way as the Earth has acquired such particles in its magnetosphere. Especially impressive are the "radio storms" lasting up to two hours in the meter-wave range. These consist of individual radiation surges, each lasting only a fraction of a second and emitting as many radio waves as thousands of millions of terrestrial thunderbolts.

The largest planet of the solar system is, accordingly, a major "jamming station" in the interplanetary area which, with its crackling and roaring, far outstrips the gentle murmur of the other planets. The radio wave commotion should be picked up even outside of the borders of the planetary system. However, Jupiter is not the only outer planet which behaves extravagantly, and its sister planets, too, have their idiosyncrasies. The most striking of these is the ring system that makes Saturn the gem of the solar system. It consists of innumerable tiny particles surrounding the planet exactly around the equator and reflecting the light of the Sun.

Especially whimsical is the behavior of Uranus (discovered in 1781); since its axis of rotation lies almost along its orbital plane, at times it practically rolls around its orbit. Neptune, too, famed for the dramatic story of its discovery, has the chance of coming up with a first-rate curiosity. If the hypothesis subscribed to by many planetary researchers and experts in celestial mechanics should prove correct, Neptune has provided the solar system with nothing less than the ninth planet. The orbit of Pluto, the outermost planet, which is totally different than those usual in the planetary system of the Sun, cuts across that of Neptune. The theory that Pluto is a moon of Neptune which has escaped has something to be said for it.

Asteroids—"celestial vermin"

The system of the great planets is by no means all there is in the solar system. Although the major part of the condensed matter in the region around the Sun is indeed concentrated in these bodies, the small objects found in the interplanetary area and beyond should not be overlooked. These "celestial vermin" play a major role in cosmogonic questions. It seems that these small objects, which consist of Earth-like material, include most of the small planets, called asteroids or planetoids, which, in their thousands, fill up the gap mentioned between Mars and Jupiter, but which are also found in other regions. The largest of them are spherical bodies of less than a thousand kilometers (600 miles) in diameter while the smallest ones are meter-sized irregularly shaped boulders —obviously debris resulting from the collisions of larger objects. From time to time, these fragments also fall on other celestial bodies.

On the Earth, these messengers of the solar system are eagerly sought by researchers so that they can be examined in the laboratory. It is this circumstance which enables astronomers to be quite well-informed about the small bodies of the planetary system in particular. The composition of meteorites indicates that most of them already have a complicated history behind them, only excerpts of which can be reconstructed. Meteorite researchers now agree that this material is not identical with the original condensate of the solar system. Nevertheless, it is certain that it has undergone fewer transformations and processes of differentiation than the material found on the surfaces of the large planets and satellites.

The substance of the comets, which, apart from the planets, are the most impressive phenomena in the solar system, is in a certain sense "Jupiter-like." They consist, very broadly speaking, of ice (i.e., frozen water, ammonia, methane, carbon dioxide), and of impurities in the form of solid particles of less volatile material frozen in the ice, and so on. If a ball of "dirty snow" of this kind, on its markedly elliptical

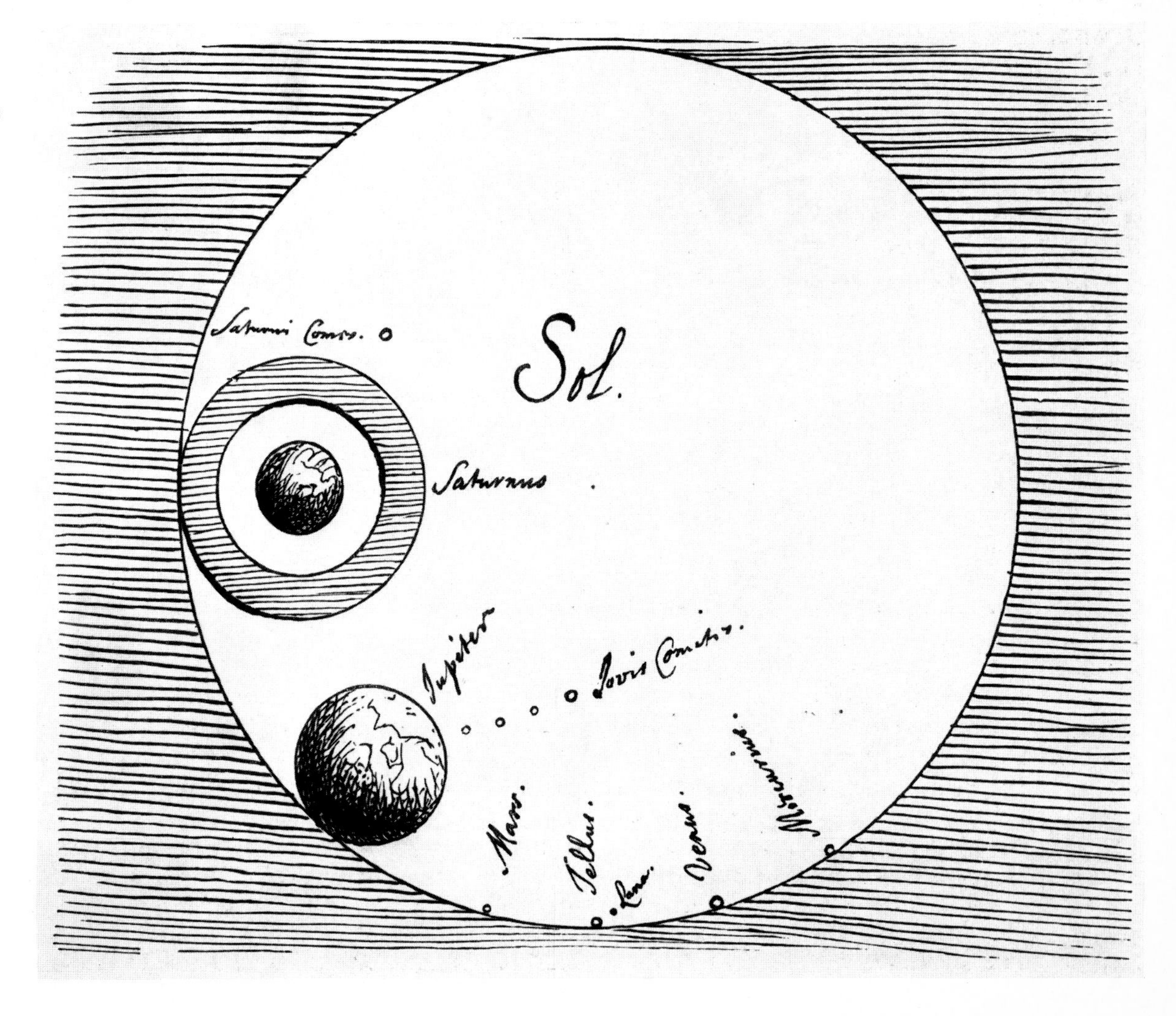

136 As long ago as the middle of the Seventeenth Century, correct assessments had been made about the dimensions of the bodies of the planetary system relative to the Sun. The drawing is the work of the Dutch astronomer Christian Huygens.

137 A good idea of the dimensions in the planetary system can be given in the following manner. If the sphere of the Television Tower in Berlin with its diameter of 32 meters (105 feet) were to represent the Sun, the outermost planet Pluto would orbit it at a distance of 140 kilometers (87 miles) and, if it were to the same scale, would be the size of a football.

138 The appearance of the giant planet Jupiter is totally dependent on what is happening in its mighty atmosphere. Its coloring and the structure of the cloud systems which form the characteristic belts are constantly changing. The so-called Great Red Spot on the right in the picture on the southern hemisphere of the planet is probably a turbulence phenomenon.

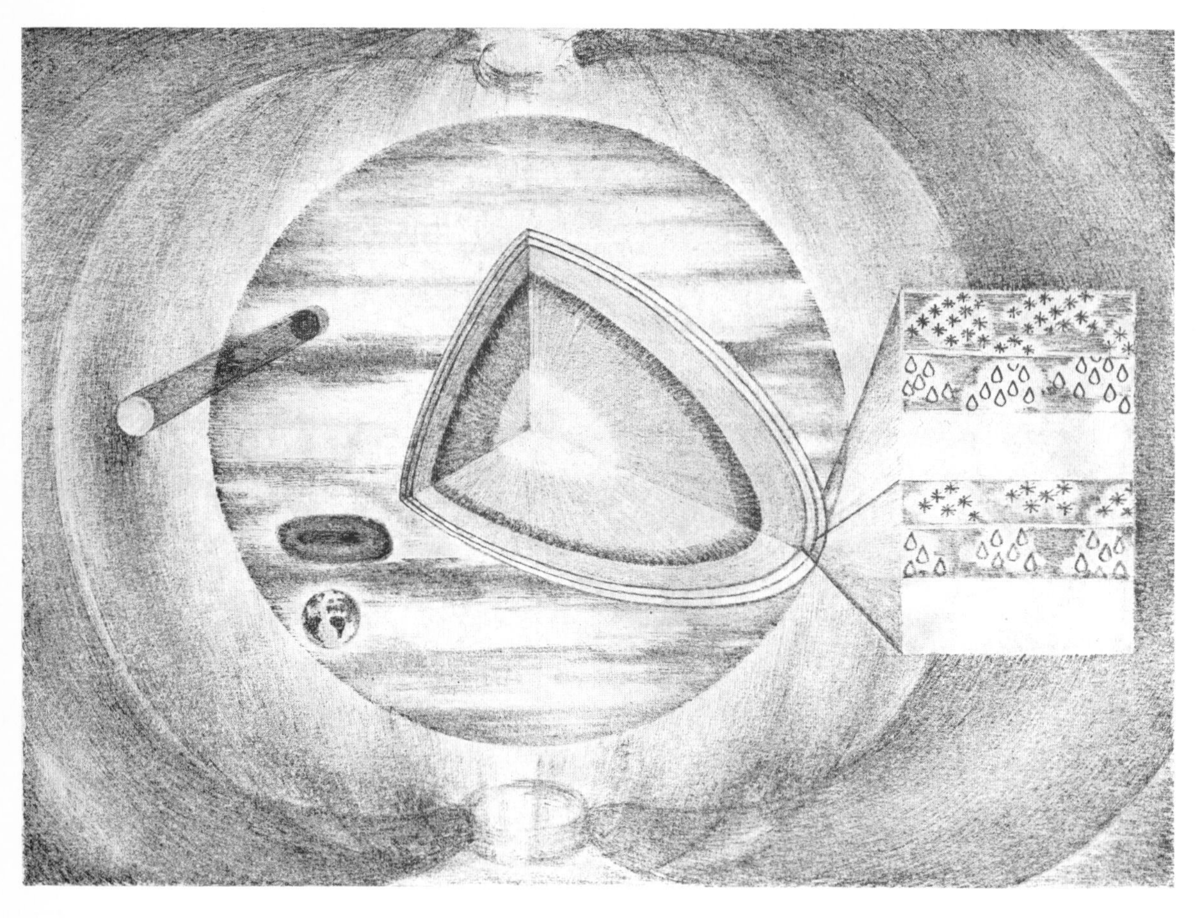

139 According to calculations based on theoretical models, the great sphere of Jupiter, noticeably flattened at the poles, has the following structure. Up to 80 percent of the radius is occupied by a core of metallic hydrogen which may possibly enclose a smaller core of heavy elements. Around this, there is a mantle of hydrogen in the normal solid state and above this a layer of the same element in the liquid state, which joins up with the atmosphere of the planet. For the latter, a model of two cloud layers has been taken as a basis, this theory being supported by the double shadows of the satellites of Jupiter which are occasionally observed. The upper layer consists of clouds of ammonia, those at the very top being composed of crystals, followed by clouds of droplets with a layer of cloudless ammonia vapor underneath. In similar fashion, the lower layer can be divided into three sublayers of ice-crystal clouds, clouds of water droplets and water vapor. The planet is enclosed by a mighty radiation belt which makes its presence felt by its radio radiation. The drawing of the Earth on the picture conveys an idea of the size of the Great Red Spot.

140 On January 7, 1610, Galileo Galilei discovered the satellites of Jupiter with the aid of a telescope which he had himself constructed, and he recorded this discovery in his diary of observations.

orbit, comes closer to the Sun than, say, the planet Mars, the frozen substances begin to vaporize. Large amounts of gas and dust particles are released from the celestial body which is only a few kilometers in size. Under the action of solar radiation, the gases become luminescent and produce the familiar tail.
Some of the comets, which every year light up in the vicinity of the Sun and thus become visible, move as short-period comets within the actual area of the planetary system, but others have orbits whose most distant points from the Sun are about one or two light-years away and thus cover interstellar distances. Estimates indicate that a cloud of some 100,000 million comets encloses the actual planetary system of the Sun. Disturbances by the closest stars cause some members of this cloud which move on very wide orbits around the Sun to break away from it and be lost in interstellar Space. On the other hand, such disturbances can turn a comet to the inner regions of the solar system.
Some scientists believe that the comets are condensation products from the early days of the planetary system. According to modern thinking on the subject, the Sun developed in the interior of a dense "cocoon" of interstellar matter. In the outer regions of this "cocoon" even volatile substances were able to condense from the dense, cool gas and develop to the size of cometary nuclei, while in the hot inner regions, in the vicinity of the emergent Sun, only the less volatile components condensed. It was these components that supplied the structural material of the terrestrial planets in particular. The details of the planet-forming process in the interior of the "cocoon" are still unknown.

Cosmic dust

In our cosmic homeland very small bodies still exist—namely micrometeorites and interplanetary dust. Instruments on board artificial Earth satellites and space probes can easily record these particles. This kind of particle has even been collected by sounding rockets and on Moon flights and brought back to Earth. It seems that the entire inner planetary system is enclosed in a flat cloud of dust which, through the scattering of light from the Sun, can be seen as a slightly brighter area, zodiacal light, near the principal plane of the planetary system. These particles, which can be as small as the wavelength of light in diameter, probably originate from disintegration processes, but most of them come from comets when they pass the Sun.

Finely dispersed cosmic dust is not the exclusive characteristic of the solar system, however, since it is a widespread form of condensed matter in Space. Concentrated in clouds of many light-years in diameter, it is found in the distant parts of the Milky Way system as a component of interstellar matter.

As long ago as the Eighteenth Century, it was noticed by William Herschel that dark "holes" can often be seen in the silvery ribbon of the Milky Way. In reality, however, it is not stars which are missing here, but dark, light-absorbing clouds situated in front of the general background of stars. Only a few decades ago this light-absorbing material was confirmed as dust and not as large fragments, for instance. Accurate quantitative investigations have shown that apart from the pronounced extinction of light in individual areas of the Milky Way due to dense complexes of dust, there is a general but much less evident extinction for almost all the more distant stars in the vicinity of the principal plane of the Milky Way system.

Fortunately, the general interstellar extinction has a varying effect on the light from the stars on the different wavelengths. The greater the amount of light that is absorbed, the "redder" the star appears, indicating to the astronomer that the apparently low magnitude is due not only to its great distance but also to the effect of interstellar dust.

From the investigation of interstellar extinction, an idea has been gained of the distribution of dust throughout the Milky Way system. It seems that the dust clouds prefer to gather in a layer within the galactic disk. Since the Sun and, therefore, terrestrial astronomers, too, are within this dust-zone—which is about 600 light-years thick—the interstellar dust prevents them from seeing outwards when their line of sight lies within this extinction layer. On the other hand, there is no interference when the line of sight is at right angles to the plane of the Milky Way. This circumstance is apparent in the sky from the fact that the distant stellar systems, the extragalactic nebulae are "missing" in the area of the Milky Way whereas they can be seen in other parts of the sky. The situation is similar in other stellar systems, too. The chains of dust clouds clearly range along the spiral arms and those systems of which we have an exact "edge-on" view are mostly characterized by a dark stripe marking the principal plane.

If a bright star happens to be in a cloud of interstellar dust, the dust then becomes visible from the light reflected by the particles. The best-known reflection nebula of this kind surrounds the Pleiades (the Seven Sisters). In many respects, even the zodiacal cloud in the solar system might be explained as a reflection nebula of the Sun.

From the investigations carried out so far, we can conclude that the dust particles are not much bigger than the wavelength of light. Their chemical composition and their origin are two of the astronomical problems which have not yet been solved. Nevertheless, cosmic dust is of great cosmogonic importance as a component of the matter in the Universe. If stars originate from interstellar matter, the first act in this cosmogonic drama is that a cloud of interstellar gas begins to shrink from the attraction exerted by its own mass. However, the usual gas clouds in the Milky Way system show little inclination to turn into stars spontaneously, but the particles of dust embedded in the clouds have the interesting quality of cooling gas. They thus make it easier for a cloud of gas to disintegrate into stars and

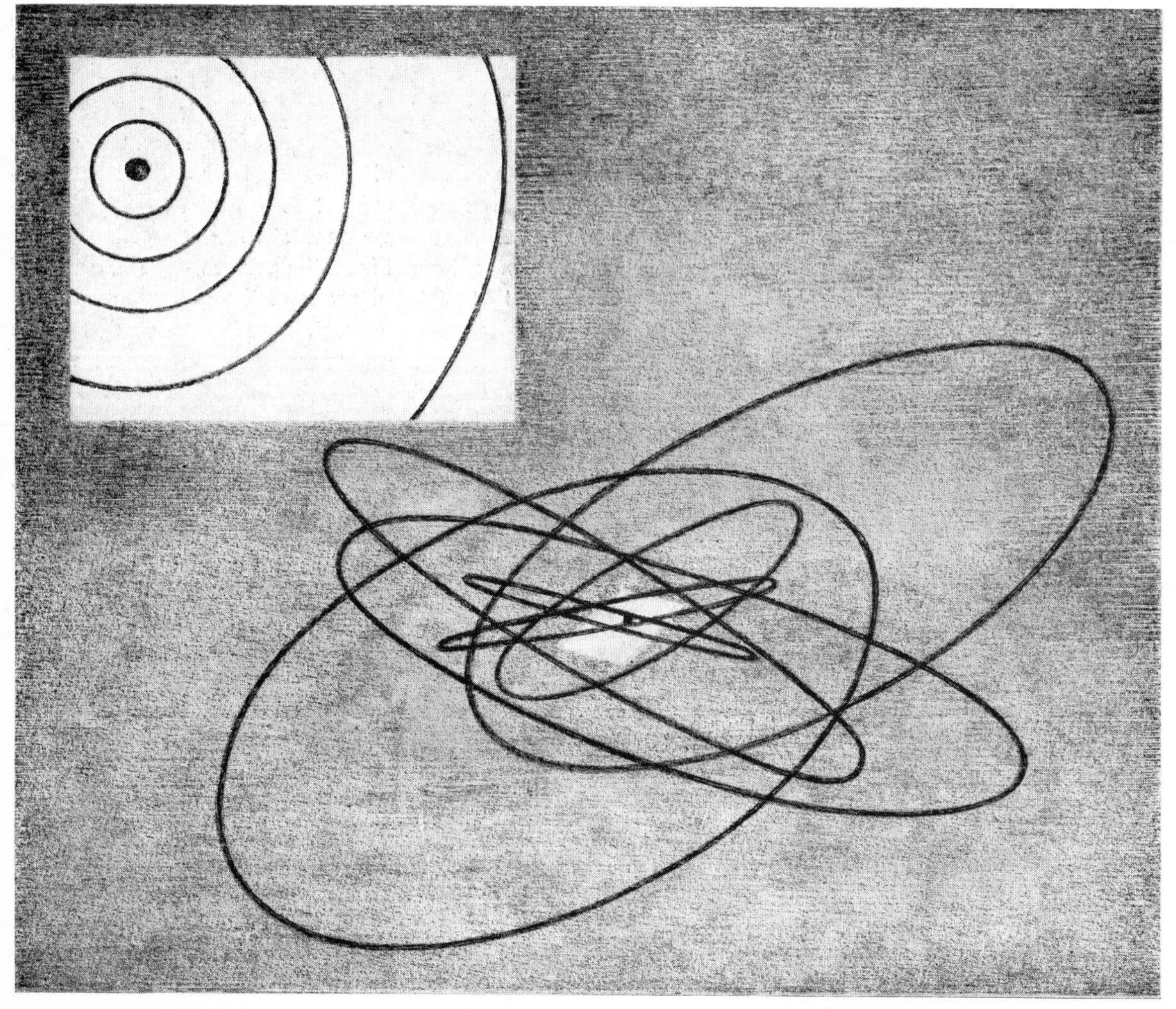

141 The five inner satellites of Jupiter, including the four discovered by Galileo, are very symmetrically arranged and are wrapped in a cloud of moons with orbits of widely varying types. The drawing is based on a model of the orbits of Jupiter's satellites by Owen Gingerich.

142 Photographs of Saturn with its decorative rings are among the most beautiful of astronomical "subjects." Saturn does not possess such striking bands as Jupiter. Saturn's rings consist of several separate zones which can be explained by celestial mechanics.

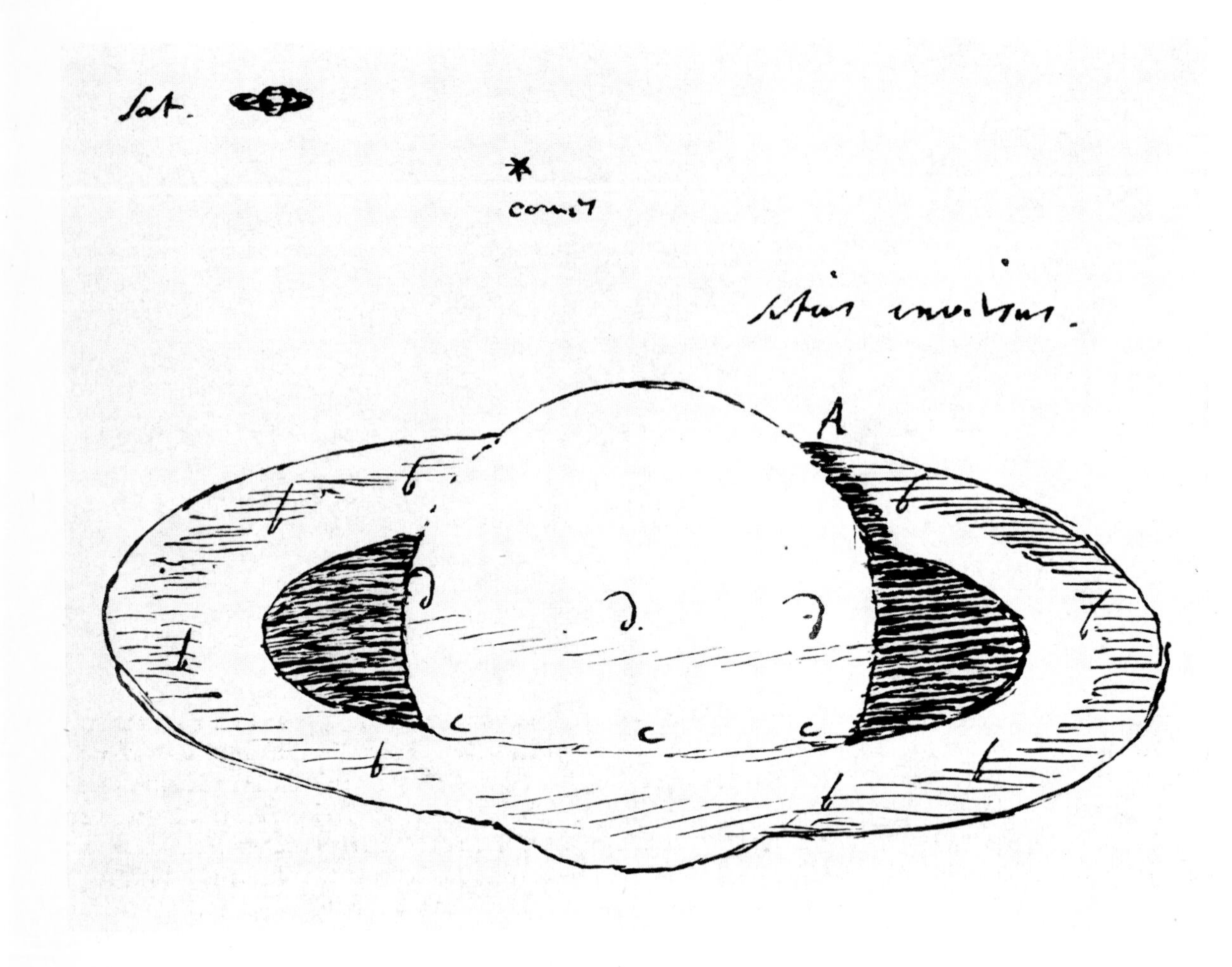

143 Since the Seventeenth Century, the planet Saturn has been regarded as the most rewarding object for telescope observations. This picture was drawn by Huygens in 1675. At the upper left, there is a sketch of Saturn with Titan, the largest satellite of Saturn, discovered by Huygens twenty years earlier.

consequently promote stellar evolution. This is, in turn, an essential first step in the emergence of planetary systems.

Alien planetary systems

What, in the past, was a purely speculative question, i.e., whether alien planetary systems existed and how such systems develop in general, has now become a serious problem engaging the attention of many researchers. The basis of their work is the various information obtained by observations. It is a striking fact, for instance, that very young stars are frequently "packed" in envelopes of dust. In many cases, the whole of the radiation of an emerging star is apparently absorbed by the envelope of dust around it and emitted as thermal radiation in the infrared range by the particles of dust. Other objects are characterized by a greater or lesser degree of extinction or have a dense reflection nebula. As intimated already, our Sun was enclosed in such a "cocoon" of gas and dust which favored the development of the planetary system. Dense and cool envelopes such as these form ideal conditions for the condensation of a gas to liquids and solids. For this reason, a number of astronomers now consider that planetary systems are an almost inevitable by-product in stellar evolution and that the surroundings of emergent stars, the circumstellar region, are the most important breeding places of condensed matter in general.

These considerations are confirmed by observations of the stars nearest the Sun, because there are signs that some of them are orbited by dark companions of low mass. Bodies with masses of the magnitude of Jupiter must, of course, initially be interpreted as planets, and it is probable that the transition to black dwarfs only takes place with somewhat larger masses. For the time being, it is not possible actually to detect these dark planetary companions since the central star of the system outshines everything else with its very much greater brightness. The existence of planets can only be detected by the slight influence they exercise on how the star moves. The star oscillates around a central position in the sky, in keeping with the orbital time of its large planets, its proper motion being serpentine. Under favorable conditions, however, the minimum fluctuations in the case of nearby stars can be detected by the sophisticated measuring methods of astrometry.

Of the proposals made so far for obtaining evidence of other planetary systems, the astrometric approach offers the most chances of success. The positive signs which have been picked up so far are only a first step, however. If, despite the low probability factor involved, signs of the existence of planetary systems are found among the direct neighbors of the Sun in the Milky Way system, this cannot be a question of a rare occurrence. Another significant fact is that all the cooler main-sequence stars, unlike the very young stars, revolve just as slowly as the Sun, exactly as if, like the latter, they had lost the angular momentum acquired in the contraction of the interstellar cloud to circumstellar material.

This optimistic outlook ends our "tour" through the condensed matter in Space which began with a special planetary system, that of the fixed star known to us as the Sun. An "excursion" into the sphere of cosmic dust brought us into contact with the problem of how stars originate. The processes taking place in the course of stellar evolution reveal the formation of planetary systems as a completely normal stage in cosmic evolution. Finally, observations have produced the first indications that in all likelihood alien planetary systems already exist in the region around the Sun.

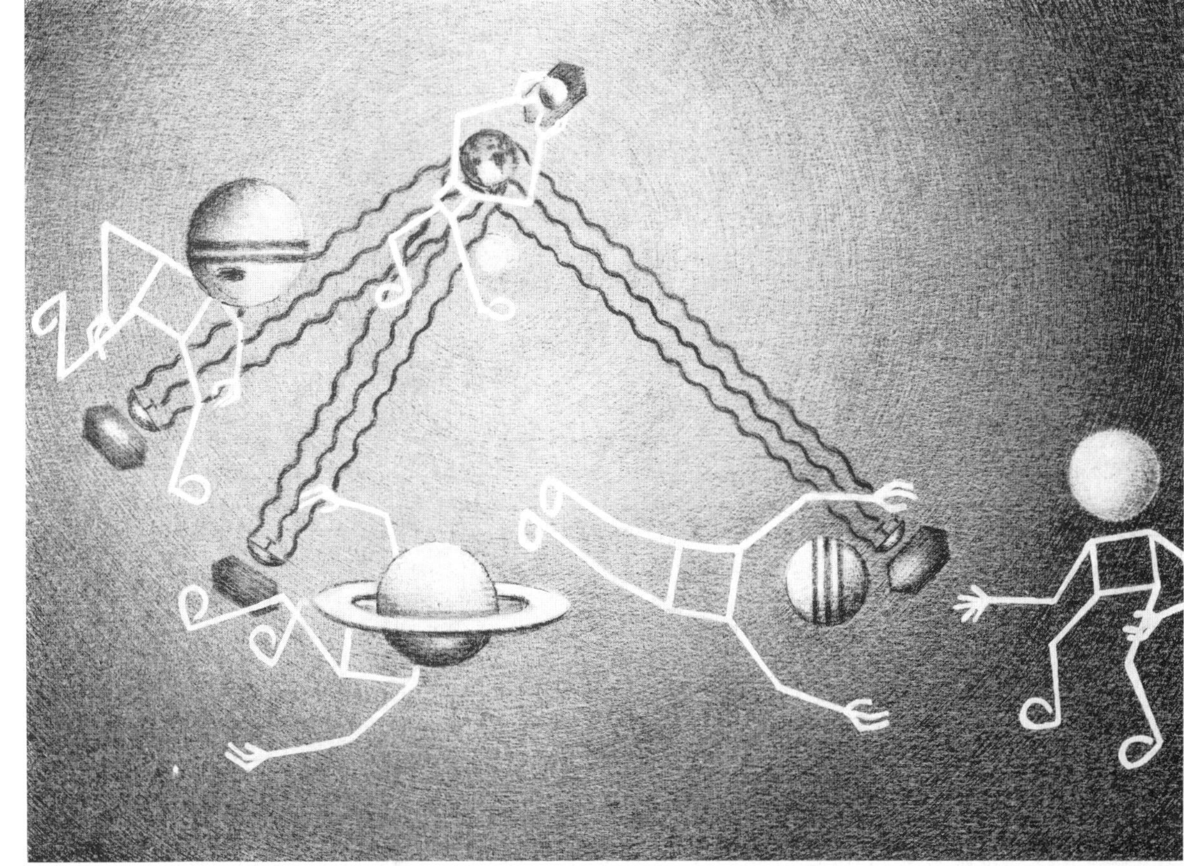

144 Between 1977 and 1979, there will be an exceptionally favorable planetary configuration for launching space probes on a "Grand Tour." The probes will pass the giant planets one by one, will gain additional acceleration from each of them and will be "passed on" to the next planet. Due to these circumstances, a time-saving flight of this kind may be said to resemble a successful combination of passes on the football field, the planets representing the players and the space probe the ball.

145 This contemporary woodcut depicts a meteorite which fell in 1492 and from which material still survives.

146a—d Unlike terrestrial rock, the largest group of meteoritic stones, chondrites, are characterized by small glassy spherules embedded in a fine-grained matrix of light grey to black in color. The largest of the chondrites shown (a), which fell in 1819 near Pohlitz, in Thuringia, is clearly marked by a thin black fused edge from its flight through the Earth's atmosphere. The dark spherules can easily be identified on the stone meteorite (b) which fell near Bjurböle, Finland, in 1899. The polished section of the dark chondrite (d, bottom) which came down in 1845 at Barratta in Australia clearly reveals the light-colored spherules. The photograph (c) shows a part of the meteorite which fell at Siena, Italy, in 1794. It consists of components of completely different composition firmly cemented to each other.

Extraterrestrial life

A review of the solid bodies and liquids in the Universe would be extremely disappointing, however, if no account was taken of a fascinating attendant phenomenon of the condensed state: the existence of life. Apart from the usual astronomical motives, the quest for other planetary systems is especially interesting because it is closely associated with the question of the existence of extraterrestrial living beings or even of alien worlds populated by rational creatures. Since ancient times, eminent thinkers have concerned themselves with the problem of extraterrestrial life, and people interested in astronomy have always wanted to have an answer to it. It is said that the circulation of the New York Sun increased by five times when a practical joker published a series of apparently serious "scientific reports" in 1835 describing the inhabitants of the Moon whom Sir John Herschel (the son of the illustrious astronomer William Herschel already mentioned a number of times here) was supposed to have discovered with a new giant telescope at the Cape of Good Hope. The idea of alien civilizations, of course, is an essential feature of the science fiction literature of the Space Age.

For the researcher, the glittering concept of life in Space, which exerts such an influence on the public imagination that every prematurely or jocularly reported discovery in this field is uncritically accepted, is really a series of sober questions and equally sober answers. The first question is usually that of the basic principle of terrestrial life, since this is the only type of life which up to now can be scientifically investigated. The second question immediately follows from this, that is, under what conditions were these living beings able to emerge. The first question can be addressed to the biologists, but an answer to the second can only be expected as the result of intensive interdisciplinary collaboration which, of course, is typical for the entire field of studies of condensed cosmic matter. Not only are biological and geological facts of importance here, but account must also be taken of the findings of astronomical research. Finally, there are also purely astronomical questions, concerning, for instance, the areas in Space which possess favorable conditions for the emergence of life and how evidence may be obtained to show whether life really has emerged there.

If these questions are considered one by one, it is first necessary to take a look at biochemistry. Even though there is a confusing variety of detail in the forms, colors and courses of development in the sphere of terrestrial life, there is nevertheless a large degree of agreement between the chemical processes in the smallest units of the organisms—the cells.

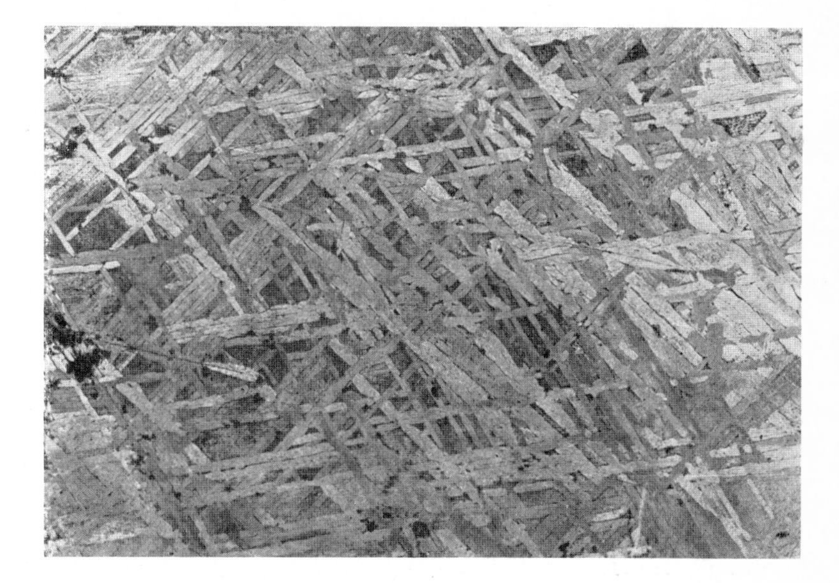

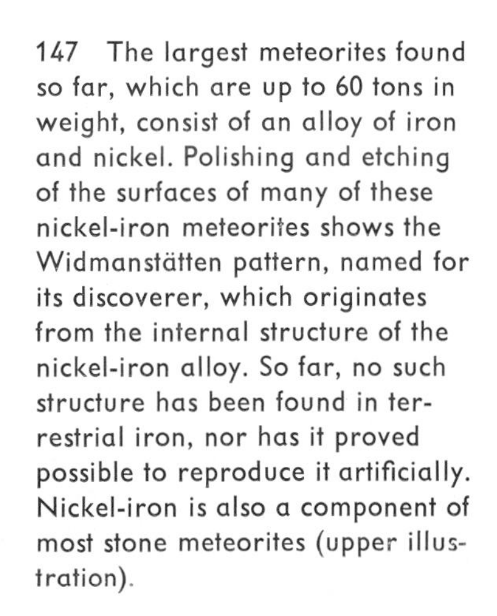

147 The largest meteorites found so far, which are up to 60 tons in weight, consist of an alloy of iron and nickel. Polishing and etching of the surfaces of many of these nickel-iron meteorites shows the Widmanstätten pattern, named for its discoverer, which originates from the internal structure of the nickel-iron alloy. So far, no such structure has been found in terrestrial iron, nor has it proved possible to reproduce it artificially. Nickel-iron is also a component of most stone meteorites (upper illustration).

148 The strangest of the rocks found on the Earth are the tectites. In the past, they were thought to be "glass meteorites," but many meteorite researchers now assume that they are terrestrial rocks melted and modified by the impact of a meteorite. The pieces illustrated come from an area in Czechoslovakia where many have been found; they belong to the subgroup of the moldavites (lower illustration).

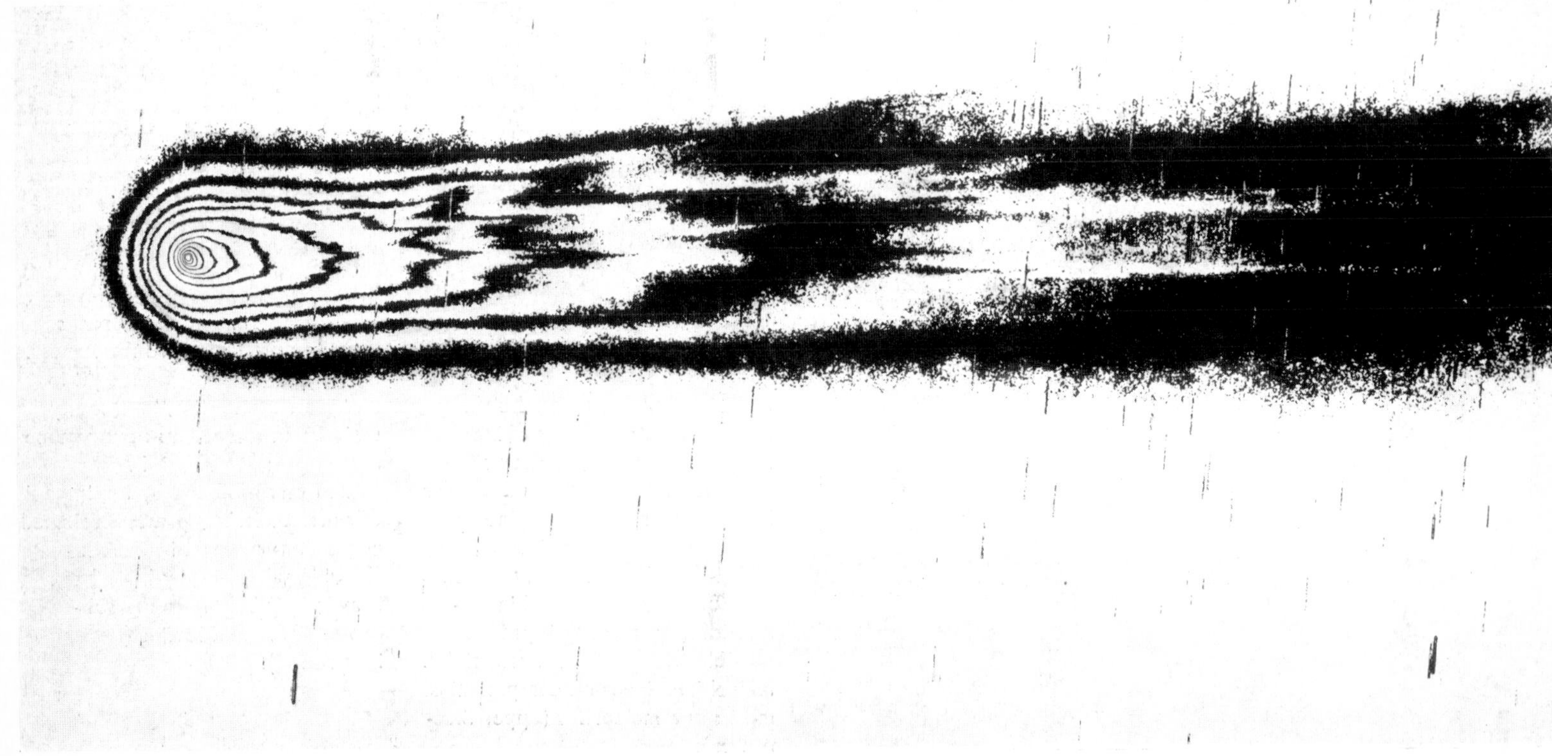

149 a, b Comet Bennett discovered in 1969 was the most impressive comet of recent years (a). The equidensity method provides a good impression of the brightness distribution in the head of the comet (b).

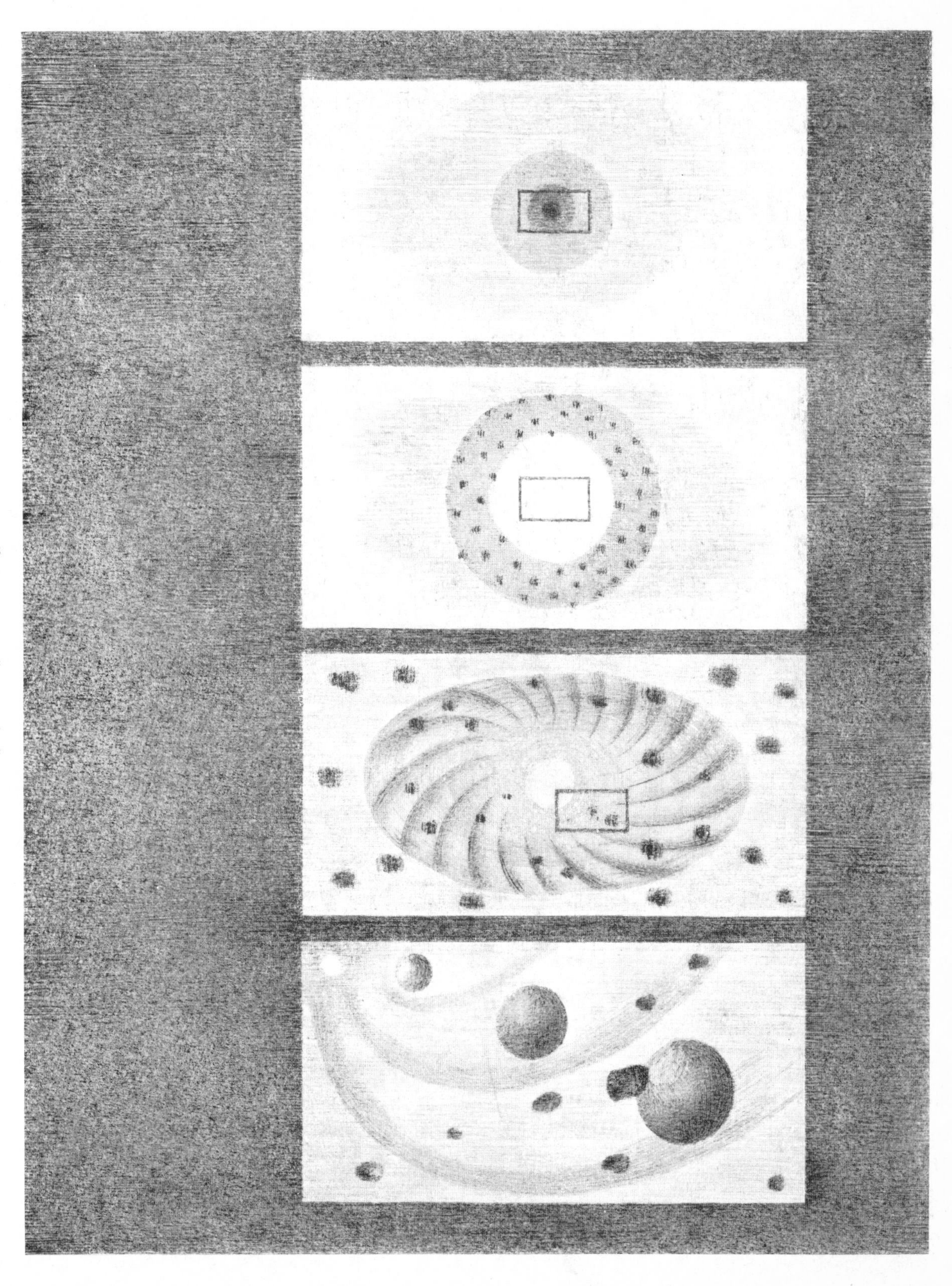

150 On the basis of modern theory, it may be imagined that the planetary system formed in the following manner: 1. An interstellar complex of gas and dust contracts due to the self-gravitation of its matter and, finally, suffers a collapse (a). 2. Due to the powerful contraction, the temperature in the interior of this "primordial nebula" rises, and a protostar is formed, enclosed by cooler layers in which numerous chemical reactions take place and gas condenses (b). 3. The protostar rotates faster and faster as it contracts and, with its powerful magnetic fields, begins to take the surrounding material with it. The protostar itself is continuously braked but the "primordial nebula" rotates with increasing speed and flattens out into a disk (c). 4. From the condensation products present in the dense disk the planets are formed (d).

One stage further, in the sphere of the molecules essential for vital processes, there is even complete uniformity and monotony. Energy and information are always transmitted in the same manner, and water is the principal component of every cell. The cell components controlling the actual vital processes, e.g., metabolism and reproduction, are macromolecules which the element carbon forms with hydrogen, oxygen, nitrogen, phosphorus and some other elements.

Organic building blocks

The most important molecule is deoxyribonucleic acid, DNA, a threadlike giant molecule about a millimeter long in the nucleus of the cell. DNA has two fundamental characteristics: it can make an identical reproduction, an exact copy of itself, and it stores in its structure all the information necessary for the synthesis of the proteins needed by and characteristic of the cell in question. DNA is thus the genetic information carrier that contains the encoded struc-

151 a, b Dark structures in extragalactic stellar systems reveal the distribution of dust as a whole and show that dust plays an important part in the life of a galaxy. There are strips of dust in the bar of the barred spiral NGC1300 (a) while in the "nebula" NGC891 (b), as seen from the edge, the dust zone can be seen near the plane of symmetry.

tural plan by which a cell can produce an identical daugther cell and, in the final analysis, by which a living being can pass on its characteristics to its descendants.
Terrestrial life is thus based on water and high-molecular carbon compounds, one of which, DNA, has the special capability of acting as a genetic information medium. Since carbon occupies a unique position in the periodic system of the elements—with the possible exception of silicon—on account of its ability to form the most complex molecules, the idea that precisely this element might play a major role in the emergence of life in other parts of the Universe as well is not at all unfounded.
The emergence of life of the kind existing on our own planet assumes an environment in which the physical conditions remain constant over a long period of time. In astronomical terms this means that life could only be expected to emerge in planetary systems associated with cool main-sequence stars which, like the Sun, evolve fairly slowly. Since most of the stars in the Milky Way system belong to this group, this scarcely imposes any restriction. More exacting is the requirement that only those planets can be considered on which water can exist in the liquid form. These must fall within a certain zone around every star which is sometimes designated as the ecosphere. The stipulation relating to carbon chemistry narrows the environment still further.
It is now generally accepted that the starting point of terrestrial life was the chemically very active original atmosphere of our planet which was presumably similar to the atmosphere of the giant planets. There are even geological indications that the original terrestrial atmosphere of more than 1,600 million years ago must have possessed reducing qualities, since the oldest sedimentary rock contains minerals which could only have been formed in an environment of this kind. It has also been demonstrated by laboratory experiments that amino acids, the "building blocks" of the proteins, can be formed from methane, ammonia and water via intermediate stages such as formaldehyde and hydrocyanic acid. Formaldehyde, however, permits the development of the vital sugars which, together with phosphoric acid, form the backbone of DNA. Hydrocyanic acid, on the other hand, is important for the development of those four heterocyclic bases whose arrangement along this backbone carries the genetic information.
Regarding the purely chemical aspects, the emergence of life is, in principle, no longer a mystery, but the cybernetic side of the problem remains largely unsolved. We still do not know how the efficient biochemical regulating circuits maintaining the system came into being, how the organization of the cell emerged, or how chemical evolution developed into biological evolution. Nor is it possible for this process in its totality to be studied in a laboratory experiment since the vast periods of time probably involved in the emergence of the first organisms cannot be simulated.
Life is thus clearly linked to condensed matter. It appears that Earth-like planets moving within the ecosphere of a star are in a particularly favorable position. For the time being, the only ecosphere that astronomers can study in any detail is that of the

152 Between 1956 and 1957, Gaposhkin was the first to draw the whole of the Milky Way from visual observations. The succession of dust complexes in the Milky Way can clearly be seen in this picture.

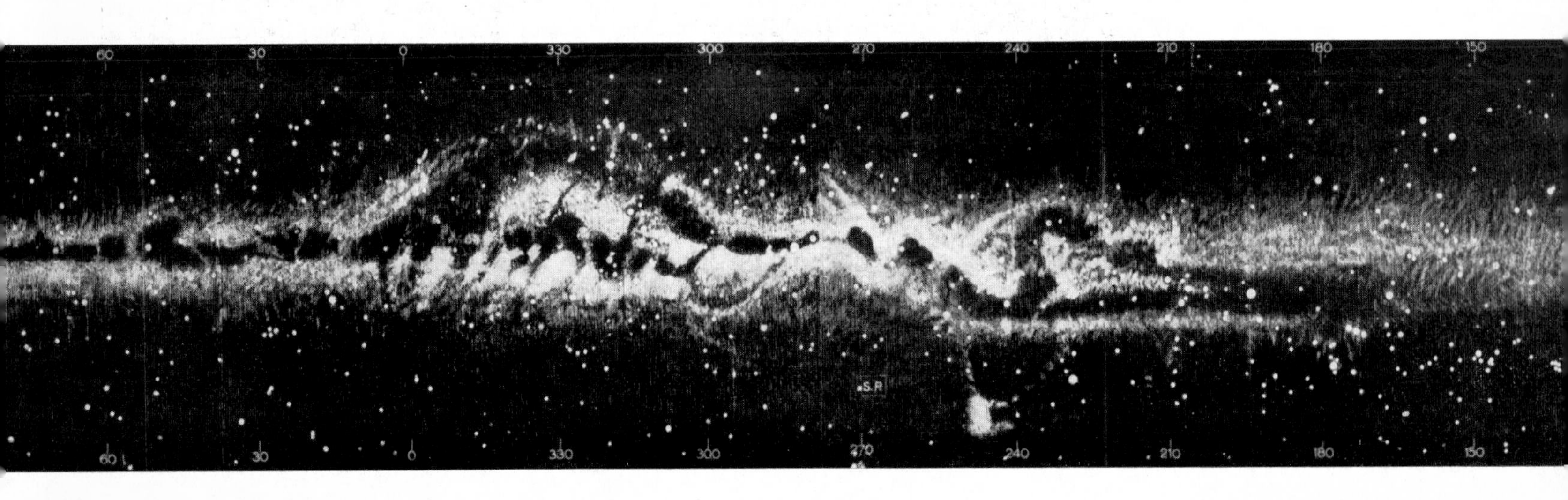

153 This gaseous nebula in Sagittarius is split into three parts by the dense strips of dust in front of it; consequently it is known as the Trifid nebula. Particles of dust also scatter the light from the hot stars and in this way produce the blue reflection nebula seen above them.

154 At the tip of this cone of dust and embedded in luminescent clouds of gas and dust, there are numerous extremely young stars which have not yet reached the main-sequence stage.

155 Apart from ragged clouds of dust, this part of the Rosette nebula in Monoceros, as shown in the picture, also contains small dark clouds of a spherical form known as globules. These are possibly the initial stages in the birth of stars.

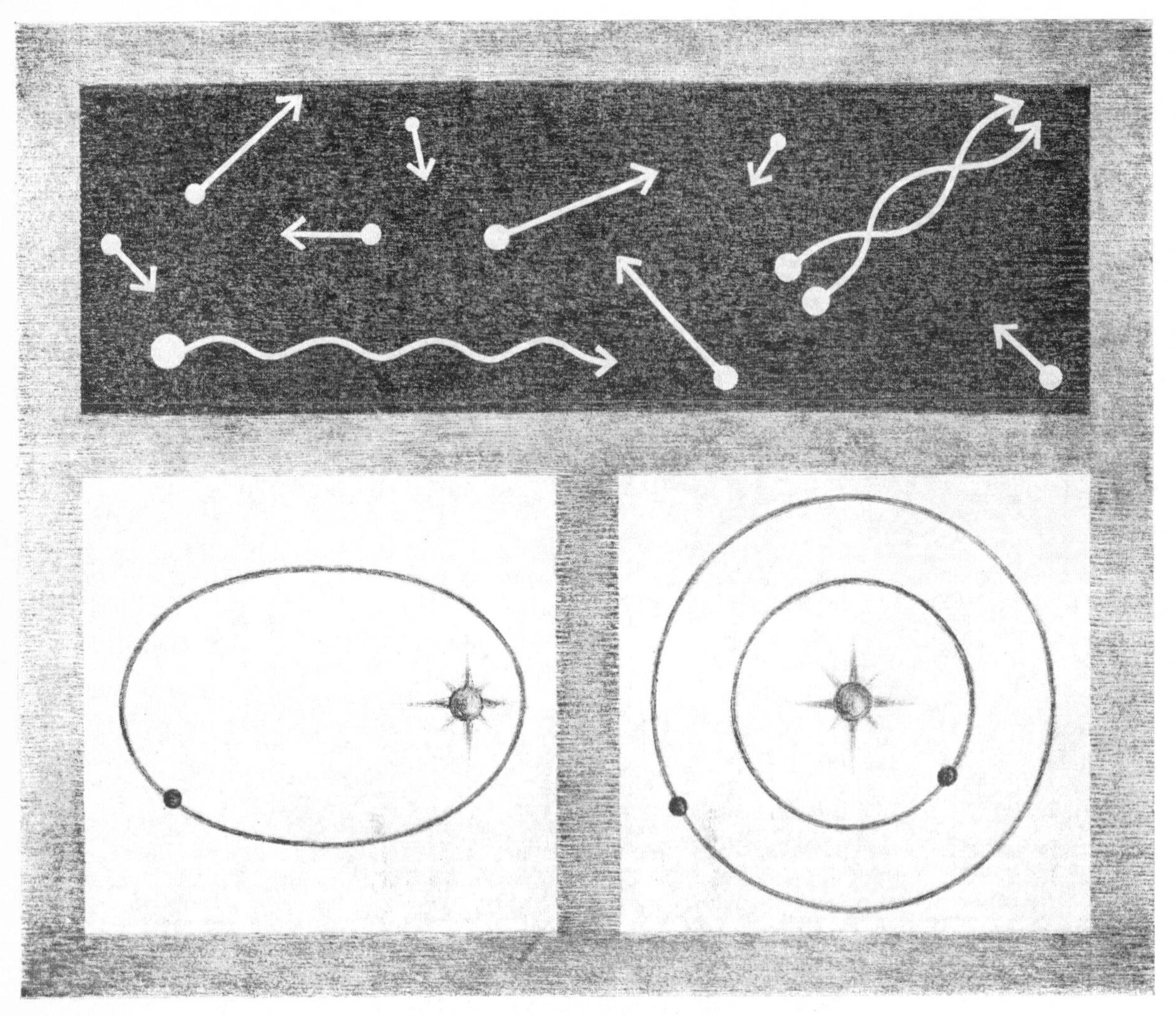

156 Single stars move in a straight line across the sky as a result of their proper motion. In the case of double stars, the proper motion of the center of gravity of the system is superimposed with the motion of the two stars around this center while stars with invisible companions attract attention by the serpentine character of their proper motion. The three types of star motion are shown in the upper part of the illustration. The analysis of the serpentine motion of Barnard's star, which is about six light-years away from the Sun, initially indicated a companion having a mass 1.7 times that of Jupiter and moving around the star in a markedly eccentric orbit (below left). Further observations imply, however, that two companions of 0.8 and 1.1 Jupiter masses on circular orbits in a single plane are also feasible (below right). This may be the first discovery of an alien planetary system.

Sun; through it moves not only the Earth but also Venus and Mars. In the other two cases, nothing that indicates life has yet come to light, although the last word still remains to be spoken. Whether the seasonal changes on Mars are actually associated with a primitive vegetation or not will become known in the coming years when Mars probes make soft-landings on the planet and carry out extensive investigations. With the aid of similar vehicles, it will also be learned whether Venus possesses cooler polar regions with more favorable conditions for life. Incidentally, the examples of Venus and Mars clearly illustrate the fact that the inclusion of a planet in the ecosphere of the star does not justify over-imaginative expectations.

Of interest for the discussion on the emergence of life in the solar system is the fact that organic substances are not at all rare. About one-hundredth of one percent of the meteorite material which has fallen on the Earth consists of carbon compounds. The proportion of organic compounds in the terrestrial biosphere in the total mass of the Earth, for instance, is a million times less. Thus the emergence of the solar system must have been associated with the production of large quantities of carbon compounds which are preserved in the small bodies of the system.

A great deal of evidence has appeared very recently to indicate that these substances are widely distributed throughout the Cosmos. Since 1969, radio astronomers have discovered a large number of organic molecules from their characteristic microwave spectrum. These molecules have been found in regions that contain interstellar dust and in which cosmogonic activities such as the emergence of stars may be expected. Interestingly enough, they include molecules of significance for the emergence of terrestrial life such as formaldehyde and hydrocyanic acid. Formaldehyde even seems to be the most widely distributed organic molecule in interstellar Space. It is probable that these compounds develop on the particles of cosmic dust, especially in the dense gas-and-dust envelopes which enclose the emergent stars. This has given a new stimulus to the discussion about the emergence and the occurrence of life in Space. The chemical basis for this is obviously even broader than hitherto assumed. Nowadays, it is generally accepted that life is neither specifically terrestrial nor unique.

The continuing quest

As in the past, there is still no direct evidence of the existence of extraterrestrial life. The coming decades will show whether there are primitive forms of life in the solar system other than on the Earth. For the other planetary systems, such evidence is completely out of the question for the time being. On the other hand, the possibility that in a distant planetary system biological evolution as on the Earth may have led to the emergence of a society of intelligent beings should not be overlooked. Such a civilization, assuming that its state of technical development is sufficiently advanced and that it has adequate sources of energy, would be able to make itself heard in the Milky Way system. Radio astronomers are thus not just playing around when they take stock of the chances of detecting signs of life from advanced civilizations and "listen" to potentially promising stars with their radio telescopes. The prospects of success of such undertakings are frequently overestimated, however. The discovery of an alien civilization and especially the establishment of contact with it will always be a very protracted task quite simply because the bridging of distances measured in light-

157 Until the last century, it was widely believed that the Moon was inhabited. In 1835, the New York Sun surprised its readers with the news that Sir John Herschel had discovered the creatures living there.

years will inevitably take years of "transmission time."

The search for intelligent beings has assumed grotesque forms in many literary products. Every mysterious feature of the Earth, every special aspect of the life of ancient civilizations not yet understood by the archaeologists, has had to do duty as "evidence" that the Earth occasionally receives visitors from the Milky Way who influence the development of mankind. In principle, from the scientific viewpoint, there is no objection to visitors from Space, but it can be expected of them that they are subject to the same natural laws as we—that they will consequently leave clear indications of their presence and furthermore that they will display a certain interest in establishing contact. However, the famous "flying saucers," the unidentified flying objects (UFOs), as they are called by the scientists who investigate these phenomena, do not at all comply with these conditions. They always vanish into thin air when examined seriously, and so there is every reason to doubt their reality.

The probability of establishing contact with the population of an inhabited planet is very slight, it is true, but it is not completely out of the question. If it actually happened, as perhaps it might in the far-off future, it would not only be of exceptional scientific interest but might well have a considerable influence on the development of mankind.

158 Four methods have so far been employed for detecting signs of intelligent life in Space. Radio telescopes have been used to "listen" to promising stars. Objects of unknown origin, such as Flying Saucers, were interpreted as spaceships of alien civilizations. Actions by visitors from Outer Space were postulated for the explanation of archeological riddles, especially in the areas of the Mayas and the Incas. In ancient manuscripts, such as the Old Testament and the Gilgamesh epic poem, indications were sought which might be explained as meeting between human beings and extraterrestrial creatures. None of these methods has yet produced any conclusive evidence.

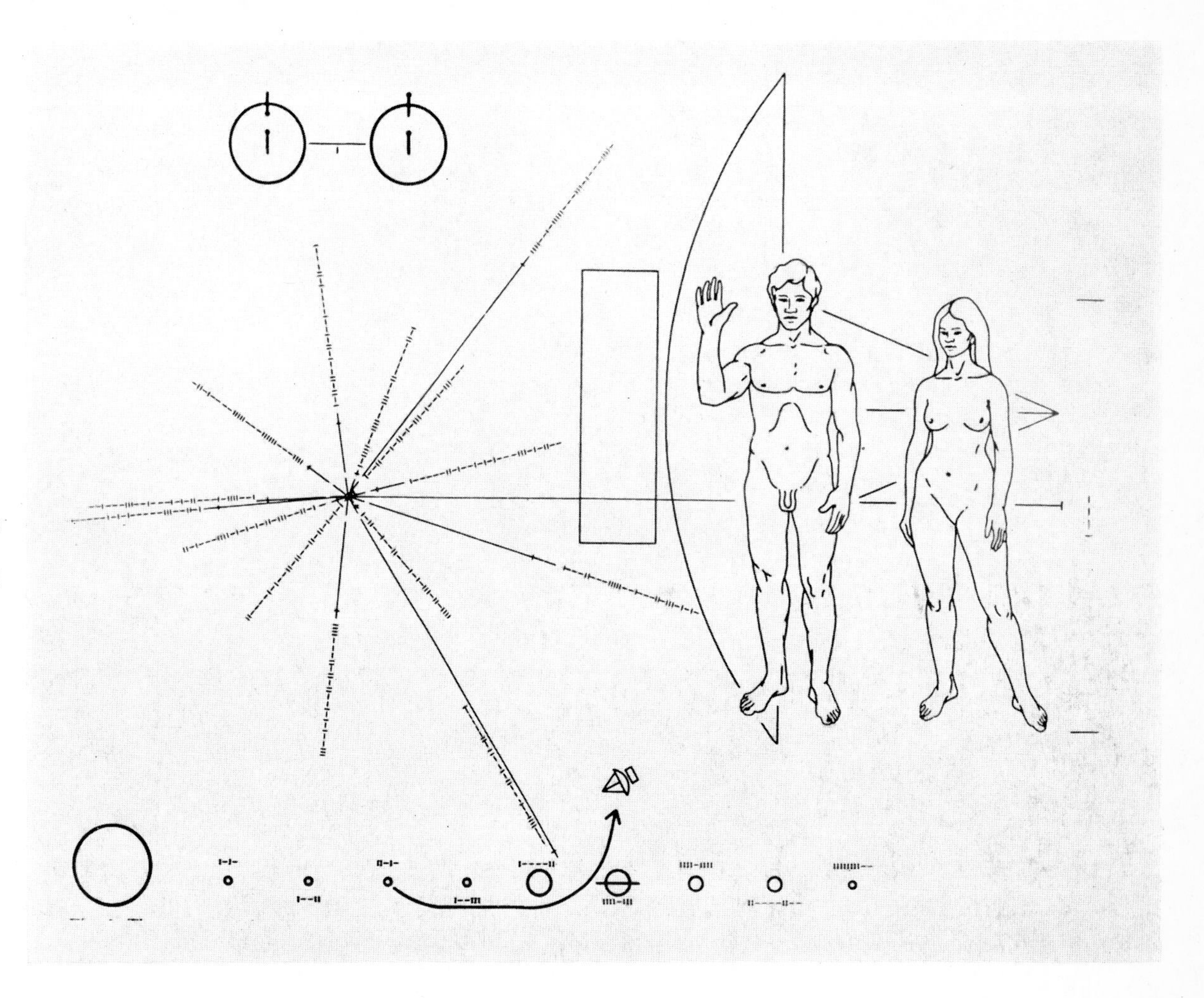

159 The Pioneer 10 space probe which was launched in 1972 will leave the solar system as the first messenger of mankind. It carries a plaque with a message to inform other intelligent beings of our existence. Apart from the illustration of two human creatures with an indication of their size, the plaque carries diagrams that clearly show the position of the Sun in the Milky Way system, the position of the Earth in the planetary system of the Sun and the course taken by the probe. All numerical data is in a binary code in which the figure 1 is defined by the transition of the hydrogen atom which causes the 21-centimeter line. This key magnitude is marked at the top left of the illustration. Living creatures with an advanced civilization and an interest in astronomy should be able to decipher this data.

Footnotes

[1] Kepler: Dioptrice. Augsburg 1611
[2] Encke: Gedächtnisrede auf Bessel. Berlin 1846
[3] Fraunhofer, Denkschr. der Königlichen Akademie der Wissenschaften zu München für 1814 und 1815. Band V, p. 193 (of 1817)
[4] Kirchhoff, Abh. der Königlichen Akademie der Wissenschaften zu Berlin aus dem Jahre 1861, p. 63
[5] G. P. Bond in a letter to W. Mitchell, in 1857. Published in: Publ. of the Astron. Soc. of the Pacific, Vol. 2, p. 300 (1890)
[6] ibid
[7] Lecture by van de Hulst. Published in: Observatory, Vol. 91, p. 55 (1971)
[8] W. Herschel in: Philosophical Transactions for the Year 1785, p. 213

Astronomical data

Basic expressions

Very large and very small numbers can be expressed conveniently in the form of powers. The number in question is arranged as the product of two numbers. One of these is usually between 1 and 10, while the other is a ten with an exponent (a whole number expressing the power, written as a small figure at the upper right of the number). This represents the magnitude of the number to be expressed. If the exponent is positive, it indicates the number of zeros which would be placed behind the one if the number were written out in full. If the exponent is negative, it indicates the number of places that the decimal would be moved to the left. For example:

$6.96 \cdot 10^5 = 6.96 \cdot 100000 = 696{,}000$

$6.96 \cdot 10^{-5} = 6.96 \cdot 0.00001 = 0.0000696$

For expressing astronomical distances, the unit known as the "light-year" is used (ly). This is the distance covered by light travelling at its speed of 300,000 km/s in a vacuum in the course of a year: $1 \text{ ly} = 9.46 \cdot 10^{12}$ km.
For expressing distances within the planetary system, the unit known as the "astronomical unit" is used (A.U.). It is the mean distance of the Earth from the Sun, i.e.,
$1 \text{ A.U.} = 1.496 \cdot 10^8$ km.

In astronomy, the scale of absolute temperature introduced by Lord Kelvin is the system most used. It is based on absolute zero. Temperatures expressed in centigrade (C) are converted to degrees Kelvin (K) by the addition of 273.

Sun

mass:	$1.99 \cdot 10^{30}$ kg or 333,000 times the mass of the Earth
radius:	$6.96 \cdot 10^5$ km
luminosity:	$3.90 \cdot 10^{23}$ kW
temperature	
(surface):	5,785 °K
(core):	$1.36 \cdot 10^7$ °K
density	
(photosphere):	approx. $1 \cdot 10^{-7}$ g/cm³
(core):	approx. 100 g/cm³
(mean):	1.41 g/cm³
rotation period	
(equator):	25.03 days (Unlike a solid body, the Sun rotates more slowly in the polar areas than at the equator.)
loss of mass	
(radiation):	$4.3 \cdot 10^9$ kg/s
(solar wind):	approx. 10^9 kg/s
solar wind:	(at the distance of the Earth): 400—500 km/s average speed
	average particle density (at the distance of the Earth): approx. 5 protons/cm³

Earth

mass:	$5.98 \cdot 10^{24}$ kg
radius:	6,371 km
mean density:	5.52 g/cm^3
rotation period:	23 h 56 min. 4 s
orbital period around the Sun:	365.2422 days (tropical year)

Moon

mass:	$7.35 \cdot 10^{22}$ kg
radius:	1,738 km
mean density:	3.35 g/cm^3
rotation period:	27.32 days (sidereal month)

Planets

If not stated otherwise, the figures in the table are expressed in units of the relevant data for the Earth.

Planet	Mean distance from Sun A.U.	Orbital period days (d) years (a)	Mass	Radius	Mean density g/cm^3	Rotation period at equator	Moon
Mercury	0.39	88 d	0.06	0.38	5.42	59 d	—
Venus	0.72	225 d	0.81	0.96	5.25	243 d	—
Mars	1.52	687 d	0.11	0.53	3.96	24 h 37 min	2
Jupiter	5.20	11.86 a	317.82	11.04	1.33	9 h 50 min	12
Saturn	9.54	29.46 a	95.11	9.47	0.68	10 h 14 min	10
Uranus	19.18	84.02 a	14.52	3.70	1.60	10.8 h	5
Neptune	30.06	164.79 a	17.22	3.50	1.65	15.8 h	2
Pluto	39.75	247.7 a	approx. 0.5			6.4 d	—

Small bodies of the solar system

Object group	Diameter	Mass	Chemical composition	Number
small planets	from $<$ 1 km to 770 km	from 10^{12} to 10^{21} kg	rock	$>$ 30,000
comets	from 1 km to 100 km	from 10^{11} to 10^{17} kg	iron conglomerate	about 10^{11}
meteorites	$<$ 3 meters	$<$ 65 t	rock, iron	
micrometeorites and dust	$< 10^{-4}$ meters	$< 10^{-9}$ kg	rock, iron	

Interstellar matter

average gas density:	$1 \cdot 10^{-24}$ g/cm^3 (about 1 hydrogen atom/cm^3)
average dust density:	approx. $1 \cdot 10^{-26}$ g/cm^3
average cloud size:	approx. 20 ly

Stars

If not stated otherwise, the figures in the table are expressed in the relevant solar units.

Object group	Surface temperature K	Mass	Radius	Luminosity	Average density g/cm³
dwarf stars (main-sequence stars)	37,800	17.5	7.5	$1.2 \cdot 10^4$	$5.6 \cdot 10^{-2}$
	14,800	6.5	4.0	$6.3 \cdot 10^2$	0.14
	9,700	3.2	2.6	$7.9 \cdot 10^1$	0.25
	5,960	1.1	1.05	1.3	1.35
	3,860	0.49	0.63	$6 \cdot 10^{-2}$	2.82
giant stars	5,400	2.5	6	$3.1 \cdot 10^1$	$1.4 \cdot 10^{-2}$
	3,500	4.5	25	$2 \cdot 10^2$	$3.5 \cdot 10^{-4}$
	2,900	5	40	$4 \cdot 10^2$	$1.1 \cdot 10^{-4}$
supergiants	6,400	12	60	$8 \cdot 10^3$	$7 \cdot 10^{-5}$
	4,000	13	$2 \cdot 10^2$	$1 \cdot 10^4$	$2.3 \cdot 10^{-6}$
	2,800	17	$5 \cdot 10^2$	$3.2 \cdot 10^4$	$1.7 \cdot 10^{-7}$
white dwarfs	10,000	0.6	$1.3 \cdot 10^{-2}$	$1 \cdot 10^{-4}$	$4 \cdot 10^5$
neutron stars	approx. $2 \cdot 10^8$	0.5	approx. 20 km	approx. 10^{-8}	approx. 10^{15}

Frequency of elements in the Cosmos

On average, for 1,000,000 atoms of hydrogen there are:

126,000 atoms of helium

9,780 atoms of oxygen

3,480 atoms of carbon

956 atoms of nitrogen

316 atoms of silicon

290 atoms of magnesium

190 atoms of iron

120 atoms of sulphur

Probable alien planetary systems

Star	Distance ly	Mass of the dark companion in Jupiter masses	Rotational period around the star years
Barnard's star	6.1	either a single companion with a mass of 1.7 or two companions with masses of 0.8 and 1.1	25 or 12 and 26
Lalande 21 185	8.2	10	8
61 Cygni	11.2	8	4.8

Milky Way system	mass:	$1.3 \cdot 10^{11}$ solar masses
	number of stars:	approx. $1 \cdot 10^{11}$
	diameter:	100,000 ly
	diameter of core:	15,000 ly
	thickness of disk:	600 ly
	luminosity:	10^{34} kW in the form of light
		10^{28} kW in the form of radio radiation
	diameter of the system of globular star clusters:	150,000 ly
	distance of Sun from center of system:	30,000 ly
	time taken for one revolution of Sun around the center:	$2.5 \cdot 10^{8}$ years
	average reciprocal spacing of the stars in the area around the Sun:	4 ly
	proportion of interstellar matter in the total mass of the Milky Way system:	2%

Typical Radio galaxy (Centaurus A = NGC 5128)	distance:	approx. $1 \cdot 10^{7}$ ly
	diameter:	approx. 10,000 ly (optically visible system)
		approx. $4 \cdot 10^{6}$ ly (emission area of radio radiation)
	mass:	approx. $1 \cdot 10^{12}$ solar masses
	luminosity:	approx. $2 \cdot 10^{34}$ kW in the form of light
		approx. $2 \cdot 10^{31}$ kW in the form of radio radiation

World of the galaxies	present average distance between two galaxies:	approx. $3 \cdot 10^{6}$ ly
	time in which the average distance would double:	approx. $7 \cdot 10^{9}$ years
	maximum observed distance of a stellar system:	approx. $4.5 \cdot 10^{9}$ ly

The Largest Instruments for Astronomical Observation

	Site	Diameter of the free aperture	Focal length of the different optical systems	Completed or inaugurated in the year
I Optical Telescopes				
Refracting Telescopes	Yerkes Observatory, Williams Bay, USA	1.02 meter	19.4 meter	1897
	Lick Observatory, Mount Hamilton, USA	0.91 meter	17.6 meter	1888
Reflecting Telescopes	Spetsyalnaya Astrofizicheskaya Observatoriya, Zelenchukskaya, USSR	6 meter	24 meter 186 meter	1972
	Hale Observatories, Mount Palomar, USA	5.08 meter	16.8 meter 81 meter 152 meter	1948
Schmidt Cameras	Karl-Schwarzschild-Observatorium, Tautenburg, GDR	1.34 meter (mirror diameter)	4 meter	1960
	Hale Observatories, Mount Palomar, USA	1.26 meter (mirror diameter)	3.1 meter	1948
II Radio Telescopes			Remarks	
	Arecibo Ionospheric Observatory, Arecibo, Puerto Rico	305 meter	Reflector is not rotable, field of view is limited to 24 degrees around the zenith	1963
	Max-Planck-Institut für Radioastronomie, Effelsberg, Federal Republic of Germany	100 meter	Reflector fully steerable	1971
	National Radio Astronomy Observatory, Greenbank, USA	91.5 meter	Reflector can only be rotated in the meridian (Transit instrument)	1962

Chronological table

3379 B.C. In Central America, first recorded total eclipse of the Moon observed.

c. 1900 B.C. Work begins on the megalithic religious structure at Stonehenge, Southern England, which was also used for astronomical observations.

c. 500 B.C. The Greek philosopher and mathematician Pythagoras assumes that the Earth and planets are spherical and teaches that they move on circular orbits around the Earth at the center of the Universe.

c. 450 B.C. Anaxagoras and Empedocles of Greece teach that the Moon shines from the reflected light of the Sun.
In the world-system of Philolaos of Croton, there is a fire at the center of the Universe, around which all the planets, including the Sun and the Earth, move.

432 B.C. Introduction of the leap year cycle for the lunisolar year by the Greek mathematician Meton.

c. 380 B.C. The Babylonians prepare the best lunar tables of Antiquity.
The Greek scholar Eudoxus develops the theory of spheres.

c. 350 B.C. Heracleides of Pontus explains the daily movement of the fixed stars by the rotation of the Earth.

c. 265 B.C. Aristarchus of Samos postulates an heliocentric world-system and attempts to calculate the distances of the Sun and Moon by using geometry.

238 B.C. Alexandrian Calendar introduced by the Decree of Canopus. It provides for an extra day every four years (intercalary day) so that the course of the year coincides with that of the seasons.

c. 200 B.C. Eratosthenes of Alexandria determines the circumference of the Earth from astronomical observations according to a method attributed to the pupils of Pythagoras.

c. 150 B.C. The great Greek astronomer Hipparchus compiles the positions and brightnesses of stars in a stellar catalog. Among other things, he discovers precession, irregularities in the movement of the Moon and the unequal length of the seasons.

46 B.C. Julius Caesar introduces the Julian Calendar. This calendar reform enabled the beginning of the year almost to coincide with the shortest day of the year, months of 30 or 31 days being counted independently of the movement of the Moon and February receiving an extra day in leap years.

7 B.C. The great conjunction of Mars, Jupiter and Saturn is observed in the East; this may have been the "Star of Bethlehem."

c. 150 On the basis of the epicyclic theory of Apollonius of Perga and the theory of spheres of Eudoxus, the Alexandrian astronomer Ptolemy postulates a geocentric system of the world and summarizes the astronomical knowledge of the Greeks in his *Almagest*.

c. 550 The abbot Dionysius Exiguus proposes that the birth of Christ be taken as the starting point of modern chronology.

1054 Chinese and Japanese astronomers observe the flaring up of a bright star in the constellation of Taurus. The remnants of this stellar explosion can still be seen as the Crab nebula.

1252 Alphonso X of Castile commissions the most famous planetary tables of the Middle Ages, compiled on the basis of the Ptolemaic system of the world.

1420 The Tartar prince Ulugh Beg erects an observatory in Samarkand and repeats the measurement of the stars in the Ptolemaic catalog.

1543 In his principal work, *De revolutionibus orbium coelestium*, Copernicus postulates the heliocentric planetary theory.

1576 On the Danish island of Hven, the foundation stone is laid of Tycho Brahe's Uraniborg observatory.

1582 Pope Gregory XIII introduces the Gregorian Calendar. The old leap year rule by which every fourth year was a leap year was amended so that those years divisible by 100, but not by 400, also counted as common years. By the non-observance of this factor, an error of 10 days had developed which was remedied by arranging for October 4, 1582, to be followed immediately by October 15, 1582.

1596 The East Frisian priest, D. Fabricius, discovers the change in light of a star, which he calls "Mira," "the wonderful," in the constellation of Cetus.

1608 Lippershey, a Dutch spectacles maker, applies for a patent for the telescope made by him.

1609 Galileo constructs his first telescopes and uses them to explore the sky. He discovers surface details on the Moon, four moons of Jupiter, the phases of Venus and sunspots. He recognizes that the Milky Way consists of individual stars.
In his work *Astronomia nova*, Kepler formulates the first two of the laws on planetary motion named for him. Kepler's third law is published in *Harmonices mundi* in 1619.

1611 In his work *Dioptrice*, Kepler sets out the basic principles of geometrical optics and describes the design of the telescope named for him.

1616 The Jesuit priest Zucchius of Italy constructs the first reflecting telescope.

1638 Fontana of Naples discovers the dark spots on Mars.

1647 With his work *Selenographia*, the astronomer Hevelius of Danzig lays the foundation of scientific lunar research.

1666 The English physicist Newton discovers that sunlight is a mixture of colored lights which can be broken up.

1676 From the eclipses of the satellites of Jupiter, the Danish astronomer Römer discovers the finite velocity of light.

1687 In his chief work, *Philosophiae naturalis principia mathematica*, Newton formulates the general law of gravitation and thus establishes celestial mechanics.

1718 The English astronomer Halley, for whom the famous comet was named, draws attention to the proper motion of various fixed stars.

1728 From astrometric observations, Bradley, later Astronomer Royal at the Greenwich Observatory in London, discovers the aberration of light.

1755 In his *Allgemeine Naturgeschichte und Theorie des Himmels*, the German philosopher Kant propounds the first scientific hypothesis on the formation of the solar system.

1781 William Herschel, one of the most important observers and telescope constructors of the Eighteenth Century, discovers the seventh planet, later named Uranus.

1794 The German physicist Chladni discovers the cosmic origin of meteorites and establishes meteoritics.

1796 The French mathematician and philosopher Laplace postulates that the planets have developed from rings of gas expelled by the Sun.

1801 Piazzi of Palermo discovers the first small planet, Ceres, in the gap between the orbits of Mars and Jupiter.

1814 The South German physicist Fraunhofer carries out spectroscopic observations and records almost 600 dark lines in the spectrum of the Sun.

1838 At almost the same time, the first measurements of the parallaxes of fixed stars are carried out by the astronomers Bessel, W. Struve and Henderson.

1839 At the suggestion of Arago, the photographer Daguerre takes a picture of the Moon—the first photograph of a celestial body.

1845 The Earl of Rosse in Ireland completes his great reflecting telescope with a metal mirror of 6 ft. (1.82 m) in diameter.

1846 From the disturbances in the orbit of Uranus, the French astronomer Leverrier predicts the existence of another planet. From the positions predicted, it is discovered by Galle and given the name of Neptune. The calculations carried out independently by the English astronomer Adams are disregarded.

1861 The German physicist Kirchhoff works out the chemical composition of a celestial body from the analysis of the solar spectrum.

1877 The Italian astronomer Schiaparelli identifies lines on the surface of Mars which he calls "canali."

1877 The American astronomer Hall discovers the two satellites of Mars.

1895 At an Irish observatory, the first attempts are made to use the photoelectric effect for astronomical purposes.

1904 At the Astrophysical Observatory, Potsdam, the existence of non-luminescent interstellar gas is discovered by Hartmann with the aid of spectroscopic equipment.

1912 In the course of her investigations of pulsating variables of the cepheid type in the Small Magellanic Cloud, Henrietta Leavitt of Harvard Observatory discovers the period-luminosity relation.

1913 On the basis of Hertzsprung's work, Russell records spectral type and absolute magnitude of stars in a diagram which is later named for these two astronomers.

1917 The 100-inch Hooker reflector (2.5 meter) of the observatory on Mount Wilson in California is completed.

1919 Working with globular star clusters at Mount Wilson Observatory, Shapley infers a vastly greater size for the Milky Way system, and places its center tens of thousands of light-years distant in the direction of the constellation Sagittarius.

1924 On photographs taken with the Hooker reflector, Hubble finds variable stars in the Andromeda nebula which demonstrate its extragalactic position.

1926 In his book, *The Internal Constitution of the Stars*, the English astronomer Eddington formulates the theory of stellar structure.

1927 Oort of Holland finds signs of the rotation of the Milky Way system and localizes its center in the direction of the constellation Sagittarius.

1929 Hubble finds proof of the general red shift in the spectra of extragalactic stellar systems as predicted on the basis of the general theory of relativity.

c. 1930 The Soviet biochemist Oparin develops the modern theory of the emergence of life on the Earth.

1930 The American astronomer Trumpler discovers the general extinction in the Milky Way system due to the presence of interstellar dust.
His countryman Tombaugh finds the planet Pluto from photographs.
The Estonian-born astronomer Schmidt at the Hamburg-Bergedorf observatory builds the first of the type of reflecting telescope with a large field of view to which his name was given.

1932 The American radio engineer Jansky discovers that the Milky Way emits radio waves.

1937 Hollerith punched card machines, the forerunners of modern electronic computers, are used for the first time for complex astronomical calculations.

1939 The German nuclear physicists Bethe and Von Weizsäcker describe a cycle of nuclear reactions which takes place in the core of stars and releases energy.

1944 The German-American Baade finds two different star populations in his investigations of the Andromeda nebula.

1948 The inauguration of the 200-inch (5 meter) reflecting telescope on Mount Palomar in California takes place. It is later named in honor of Hale, the initiator of the best-known giant telescope projects of the USA.

1949 The Englishmen Bolton, Stanley and Slee identify the first discrete radio sources with optically visible objects outside the solar system.

1951 Almost simultaneously, three radio astronomy observatories discover the radiation on the 21-cm wavelength emitted by interstellar hydrogen. This was theoretically predicted by the Dutchman Van de Hulst and the Soviet astronomer Shklovski.

1952 The American astronomers Morgan, Sharpless and Osterbrock detect parts of spiral arms in the spatial distribution of extremely young, blue stars near the Sun.
By calculations, the Americans Schwarzschild and Sandage prove that the giant red stars are formed as an inevitable consequence of stellar development.

1955 The American radio astronomers Burke and Franklin discover the powerful radio radiation of the planet Jupiter.

1957 Sputnik 1, the first artificial terrestrial satellite, is launched in the Soviet Union.
In the "Stratoscope" project, solar photographs with a new quality of detail are taken from a height of 15 miles through a reflector telescope carried by a balloon.

1958 From the measurements taken by the first terrestrial satellites and space-probes, the American physicist Van Allen identifies a radiation belt around the Earth.

1959 As the first space vehicle, the Soviet moon-probe Luna 2 lands on the Earth's satellite. Luna 3 orbits the Moon and photographs the previously unknown far side of the Moon.

1960 The first X-ray picture of the Sun is taken by a research rocket.
In the National Radio Astronomy Observatory of the USA attempts are made to identify radio signals from alien civilizations.

1961 Major Gagarin of the Soviet Air Force is the first man to orbit the Earth in the spaceship Vostok 1.

1962 The Americans Matthews and Sandage identify radio sources with star-like objects for which the name "quasar" is coined.
In the USA, OSO 1, the first solar observatory orbiting the Earth, is launched.
With the aid of a research rocket, the first non-solar X-ray source is discovered in the direction of the center of the Milky Way.

1963 From astrometric investigations of Barnard's star, the American astronomer Van de Kamp postulates the existence of planetary companions.

1965 The Mars probe Mariner 4 sends back the first close-up pictures of the surface of Mars, on which craters can be seen.
Using radio techniques, two American physicists, Penzias and Wilson, discover the cosmic background radiation which had been predicted twenty years earlier as the remainder of the hot past of the Universe.
In the course of a search for celestial sources of infrared radiation carried out in the USA, objects of low temperature are found which are explained as emergent stars.

1966 The Soviet moon-probe Luna 9 makes a soft landing on the Moon and sends back pictures and data of the surface.

1967 An English research group headed by Hewish discovers a pulsating radio source, the first pulsar.
The Soviet probe Venus 4 reaches the lower layers of the atmosphere of Venus.

1968 The astronomical satellite observatory OAO 2 starts collecting data.

1969 From observations carried out by radio astronomy, formaldehyde is identified as the first of a constantly increasing number of organic types of molecules in interstellar Space.
American astronauts Armstrong and Aldrin are the first men to land on the Moon.

1971 The Soviet space-lab Salut is launched.
The three probes Mariner 9, Mars 2 and Mars 3 become the artificial satellites of Mars. Mars 3 soft-lands instruments on the surface of the planet.
Lunokhod 1, the automatic laboratory remote-controlled from the Earth covers a distance of about 10 kilometers during its more than ten months of activity on the Moon.

1972 The six-meter reflecting telescope of the Soviet Academy of Sciences is undergoing trial operation.

1973 During their one–to–three months' stay aboard the space laboratory Skylab, three three-man crews carry out comprehensive astronomical observations (with a special view to solar physics).

Glossary

Aberration: Change in direction of light from stars caused by the movement of the Earth whose velocity in relation to the speed of light cannot be ignored.

Absorption line: Narrow, dark stripe in the spectrum of an astronomical object caused by the absorption of radiation on a certain wavelength.

Achromatic lens: Multi-element telescope lens providing good color reproduction through the combination of different grades of glass.

Asteroids: Group of small celestial bodies (also known as planetoids) moving around the Sun, mostly in the area between the orbits of Mars and Jupiter.

Astrometry: Branch of astronomy dealing with the exact determination of the positions of stars in the celestial sphere (also known as positional or spherical astronomy).

Black body radiation: Electromagnetic radiation in a cavity the black walls of which are at a certain temperature.

Celestial mechanics: Branch of astronomy dealing with the theory of movement and the determination of the orbits of celestial bodies.

Continental drift: Movement of the continents of the Earth in relation to each other over long periods of time.

Coudé focus: Place where a large telescope with extra mirrors forms a fixed image when the telescope is moved.

Culmination: Time at which a star crosses the meridian, thus achieving its greatest altitude above the horizon.

Delta-Cephei star: Pulsating variable star, named for the star Delta in the constellation Cepheus, whose variation in light is typical of a group of variable stars.

Diopter: Simple sighting instrument for determining direction, consisting of a circular aperture and cross hairs.

Ecliptic: Apparent orbit of the Sun in the sky on which, in the course of a year, it passes through the constellations of the Zodiac.

Ecosphere: Zone around a star within which, on an earth-like planet, temperatures compatible with life can develop.

Equidensity: Curve linking points of equal density on photographs of two-dimensional objects.

Equinox: Two times in the year at which the Sun in its movement along the ecliptic crosses the celestial equator, rising exactly in the east and setting exactly in the west, thus remaining visible above the horizon for exactly twelve hours.

Extinction: Reduction in light caused by small particles, such as molecules and dust grains, which absorb, disperse and refract light.

Granulation: Granular pattern of dark and bright areas on the Sun's disk.

Isotopes: Atoms of the same chemical element but having different masses in the atomic nucleus.

Luminosity: Total energy emitted by a star per unit time.

Magnetosphere: Region in the fairly close proximity of a planet in which its magnetic field is of major importance for the physical conditions.

M: Abbreviation for objects in the nebula catalog of the French astronomer Messier, e.g., M 42 for the Orion nebula.

Microwave radiation: Electromagnetic radiation with wavelengths from a few millimeters to a few centimeters.

NGC: Abbreviation for objects in the "New General Catalog of Nebulae and Clusters of Stars" of the British astronomer Dreyer, e.g., NGC 7000 for the North America nebula.

Nova: Star which, in an explosive ejection of mass, increases its luminosity by a hundred thousand times within hours and then slowly returns to its original brightness.

Parallactic movement: Distance-dependent periodic movement of the stars in the celestial sphere as a reflection of the annual movement of the Earth around the Sun.

Parallax: Angle at which the orbital radius of the Earth appears when seen from a star and whose magnitude is used as a measurement of distance.

Petrography: Description and classification of rocks.

Phase: Chronologically variable state of illumination mostly in non-self-luminous celestial bodies, e.g., the phases of the Moon or Venus.

Photosphere: Layer on the surface of a star, e.g., the Sun, from which the greater part of the observed radiation originates.

Physical parameters: Physical quantities, e.g., mass, luminosity, radius, characterizing a star as a whole.

Precession: Long-term shift in the celestial equator relative to the ecliptic as a consequence of the gyroscopic movement of the Earth's axis, expressed in a systematic change of the star coordinates.

Prominence: Stream of gas which can be seen on the edge of the Sun's disk as bright plumes and in other forms.

Protostar: Interstellar cloud of gas and dust which contracts into a star.

Pulsar: Special class of radio sources, whose radiation consists of extremely short pulses at very regular intervals.

Quasar: Class of star-like (quasi-stellar) objects mainly characterized by a marked red shift.

Radar: Abbreviation of "Radio Detecting and Ranging," i.e., the detection and measurement of the distance of objects by radio waves.

Red shift: Shift of the energy distribution in the spectra of cosmic objects, especially galaxies, toward the longwave side.

Reflection nebula: Cloud of interstellar dust illuminated by a star.

Rest mass: Mass of elementary particles in the state of rest.

Scintillation: "Dancing" and "flickering" of the stars caused by turbulent disturbances in the terrestrial atmosphere.

Seismometer: Instrument for recording vibrations of the ground.

Selenography: Description of the surface of the Moon from telescope observations.

Solar wind: Constant stream of ions and electrons emanating from the Sun.

Solstice: Two times in the year when the Sun is at its greatest or least altitude (summer solstice and winter solstice, respectively) above the horizon at noon.

Spectral class: Classification characteristic for stars with the same spectrum indicating identical surface temperature and chemical composition.

Spectrograph: Item of auxiliary equipment which breaks down stellar radiation into component spectral parts and records it photographically as a spectrum.

Supernova: Star which, in an explosive ejection of mass, increases its luminosity by ten million to a hundred million times within hours and then, in the course of months or years, returns to its original brightness.

Three-Kelvin radiation: Electromagnetic radiation detectable by radio astronomy and uniformly dispersed throughout Space. Its spectrum coincides with that of a black body at a temperature of 3° Kelvin (equivalent to –270 °C).

W-Virginis star: Pulsating variable star, named for the star W in the constellation of Virgo, whose variation in light is typical of a group of variable stars.

Zodiacal light: A faint illumination of the night sky along the Zodiac.

Acknowledgements

Academy of Sciences of the U.S.S.R., Moscow: fig. 129a
Academy of Sciences of the U.S.S.R., Moscow/Myakonky: fig. 55
American Science and Engineering, Cambridge, Mass., Solar Physics Group: fig. 82
Anton, Munich: fig. 23
Archenhold-Sternwarte, Berlin-Treptow: fig. 12a—d
Astrophysikalisches Observatorium, Potsdam: fig. 41, 46
British Museum, London: fig. 2
California Institute of Technology and Carnegie Institution of Washington/Hackercolor, Sacramento, Calif.: figs. 52, 86, 94a, 101 a, b, 102, 112b, 130, 142, 153
VEB Carl Zeiss, Jena: fig. 45
Deutsche Staatsbibliothek, Berlin: figs. 10, 11
Department of the Environment, London: fig. 15
Deutsches Museum, Munich: figs. 6, 22, 29
European Southern Observatory (ESO), Hamburg: fig. 56
Film and Picture Center of the Bergakademie Freiberg: figs. 114, 120a—d
Foto-Thoma, Benediktbeuern: figs. 36, 37
Fotoarchief Sterrewacht, Leyden: fig. 62
Fraunhofer Institut, Freiburg im Breisgau: fig. 76
Harvard College Observatory, Cambridge, Mass.: fig. 152
Harvard College Observatory/ W.O.Roberts Climax, Colo.: fig. 79a
High Altitude Observatory, Boulder, Colo.: figs. 80, 81
Karl-Schwarzschild-Observatorium, Tautenburg: figs. 47, 91b, 93a, b, 95b, 149a, 149b
Kitt Peak National Observatory, Tucson, Ariz.: fig. 138
Köhler-Kurze, Bliesheim: fig. 63
Kwasan Observatory, Kyoto: figs. 123, 125c
Leif Geiges, Staufen: fig. 42
Lick Observatory, University of California, Santa Cruz: fig. 112c
Lowell Observatory, Flagstaff, Ariz.: fig. 131
Main Astronomical Observatory, Pulkovo/Krat: fig. 78a
Mineralogisch-Petrographisches Institut der Universität Tübingen: figs. 115a, 115b, 124a, 124b
Museum für Naturkunde der Humboldt-Universität zu Berlin: figs. 146a—d, 147
Museo Nacional de Antropologia, Mexico City: fig. 7
Mount Wilson and Palomar Observatories, Pasadena: figs. 51, 53, 67, 92, 100a, 103a—e, 104b, 105, 151a, 151b, 154, 155
Národní muzeum v Praze/Slovikova: fig. 148
National Aeronautics and Space Administration, Washington: figs. 69, 70, 71, 72, 73, 75, 116a, 116b, 119, 125d, 126, 127, 129b, 133a—c, 134, 159
Novosti, Berlin: figs. 54, 68, 74, 128
Optisches Museum Jena/Linde: figs. 34, 38
Oriental Institute, Luxor: fig. 5
Royal Observatory Edinburgh: fig. 60
Royal Observatory/Campbell Harper Studios, Ltd., Edinburgh: fig. 58
Sächsische Landesbibliothek, Dresden: figs. 19, 20
Sacramento Peak Observatory AFCRL, Cambridge, Mass.: fig. 79b
Schaifers, Heidelberg: fig. 157
Science Museum, London: fig. 40
Somburg, Berlin: fig. 26
Sonnenobservatorium Kanzelhöhe/Haupt: fig. 79c
Staatliche Museen zu Berlin: figs. 3, 4, 17, 18a, 18b
Staatliche Kunstsammlungen Kassel: figs. 24, 30
Staatlicher Mathematisch-Physikalischer Salon/Deutsche Fotothek Dresden: figs. 28, 39
Sternwarte Sonneberg: fig. 95a
Sternwarte Sonneberg/C. Hoffmeister: figs. 96a, 99
U.S. Geological Survey, Washington: fig. 117
U.S. Naval Observatory, Washington: figs. 90, 94b
University Library, Glasgow: fig. 21
University Library, Leipzig: fig. 9
University Observatory, Jena: figs. 85, 91a
Urania-Verlag, Leipzig: fig. 14
Washburn Observatory of the University of Minnesota/Code, Houck: fig. 97
Wattenberg, Berlin: figs. 13a—c

Reproductions have been taken from the following books:

Bessel, Abhandlungen. 1876: fig. 35
Copernicus, De Revolutionibus. 1543: fig. 9
Flammarion, Himmelskunde für das Volk. No date: fig. 27
Fraunhofer, Denkschriften der königlichen Akademie der Wissenschaften zu München für 1814 und 1815. 1817: fig. 43
Galilei, Le opere di Galileo Galilei, Vol. 3. 1892: figs. 91c, 140
Heide, Kleine Meteoritenkunde. 1934: fig. 145
Herschel, J., Philosophical Transactions of the Royal Society London. 1833: fig. 100b
Herschel, W., Philosophical Transactions of the Royal Society London. 1785: figs. 49, 50
– Bau des Himmels. 1826: figs. 93c, 93d
Hess, Himmels- und Naturerscheinungen in Einblattdrucken des 15. bis 18. Jahrhunderts. 1911: fig. 77
Hevel, Machina Coelestis. 1673: figs. 25, 31, 32
– Selenographia. 1645: fig. 122
Huygens, Œuvres complètes de Christiaan Huygens, Vol. 15. 1925: figs. 136, 143
Littrow, Atlas des gestirnten Himmels. 1867: figs. 96b, 112c
Lohrmann, Mondcharte in 25 Sectionen und 2 Erläuterungstafeln. 1878: fig. 125b
Pooley and Kenderdine, Monthly Notices of the Royal Astronomical Society. 1968: fig. 64
Rosse, Earl of, The Scientific Transactions of the Royal Dublin Society. 1880—82: fig. 100c
Schroeter, Selenotopographische Fragmente. 1791: fig. 125a
Secchi, Die Sonne. 1872: fig. 78b